Design and Fabrication of Micro/Nano Sensors and Actuators, Volume II

Design and Fabrication of Micro/Nano Sensors and Actuators, Volume II

Editors

Weidong Wang
Yong Ruan
Zai-Fa Zhou

Basel • Beijing • Wuhan • Barcelona • Belgrade • Novi Sad • Cluj • Manchester

Weidong Wang
School of Mechano-Electronic
Engineering
Xidian University
Xi'an
China

Yong Ruan
Department of Precision
Instruments
Tsinghua University
Beijing
China

Zai-Fa Zhou
School of Electronic Science
and Technology
Southeast University
Nanjing
China

Editorial Office
MDPI AG
Grosspeteranlage 5
4052 Basel, Switzerland

This is a reprint of the Special Issue, published open access by the journal *Micromachines* (ISSN 2072-666X), freely accessible at: www.mdpi.com/journal/micromachines/special_issues/Micro_Nano_Sensors_Actuators_Volume_II.

For citation purposes, cite each article independently as indicated on the article page online and using the guide below:

Lastname, A.A.; Lastname, B.B. Article Title. *Journal Name* **Year**, *Volume Number*, Page Range.

ISBN 978-3-7258-1546-3 (Hbk)
ISBN 978-3-7258-1545-6 (PDF)
https://doi.org/10.3390/books978-3-7258-1545-6

Contents

About the Editors

Weidong Wang

Dr. Weidong Wang received his B.S., M.S., and Ph.D. degrees from Xidian University in 2000, 2003, and 2007, respectively. He is a full-time professor at the School of Mechano-Electronic Engineering, Xidian University. He is also the Director of the Research Center of Micro-Nano Systems and the Deputy Director of the CityU-Xidian Joint Lab of Micro/Nano-Manufacturing. Additionally, he holds senior membership positions in other professional societies, including the IEEE (Institute of Electrical and Electronics Engineers), the CSMNT (Chinese Society of Micro-Nano Technology), the CMES (Chinese Mechanical Engineering Society), etc. He mainly engages in research on micro/nano-electromechanical systems (MEMS/NEMS), flexible sensing, and Internet of Things (IoT) technologies.

Yong Ruan

Dr. Yong Ruan received his Ph.D degree from Peking University in 2006. From 2008 to 2020, he was a Research Fellow at the School of Department of Precision Instrument, Tsinghua University, engaging in research on micro- and nanotechnology. He has been a full-time professor at Tsinghua University since 2020. His specific areas of study include MEMS devices and key technologies, as well as the design and processing techniques for silicon-based MEMS process rules. He has been awarded funding from eight different projects, including the National Postdoctoral Fund, the Ministry of Education Doctoral Fund, the Key Laboratory Testing Fund, the Aerospace Support Fund, and the 973 Program Fund. As a key member, he has participated in major projects, such as the Ministry of Education's Major Science and Technology Research Cultivation Project. Prof. Ruan has published over 40 papers, with more than 30 indexed by SCI and EI. He has also applied for 30 Chinese invention patents, of which 20 have been granted.

Zai-Fa Zhou

Prof. Zai-Fa Zhou received PhD degrees in microelectronics and solid-state electronics from Southeast University in 2009. After graduating with the highest honor, he joined the faculty of the Department of Electronic Science and Engineering, Southeast University. He is currently a full-time professor at Southeast University of China. He was appointed as the Deputy Director and the Executive Deputy Director of the Key Laboratory of MEMS of the Ministry of Education in 2012 and 2019, respectively. His current research interests include the modeling and simulation of micro/nano-fabrication processes, the in situ extraction of material parameters for MEMS/NEMS, the collaborative design of MEMS-IC, etc. Prof. Zhou has co-authored 2 scientific books, 4 chapters in scientific books, as well as more than 120 peer-reviewed international journals and conference papers.

Preface

Microelectromechanical systems (MEMSs) are microdevices or systems that integrate microsensors, microconverters, microactuators, micromechanical structures, and micropower sources. MEMS devices have a wide range of applications in biomedical, automotive, aerospace, and communications fields, among various others. The design, optimization, performance, and application of devices are crucial to the development of modern technology. In terms of design, MEMS devices need to overcome various challenges, such as size constraints, material selection, and manufacturing processes, to achieve high integration and miniaturization. Optimizing the design and performance of MEMS devices is of great significance for improving system efficiency, reducing costs, and enhancing functionality, and it is one of the current research hotspots. The performance of MEMS devices involves sensitivity, stability, power consumption, and other aspects to meet the requirements of various applications. This Special Issue explores key issues in the design, optimization, performance, and application of MEMS devices in order to provide a reference for research and development in related fields.

Weidong Wang, Yong Ruan, and Zai-Fa Zhou
Editors

 *micromachines*

Editorial

Design and Fabrication of Micro/Nano Sensors and Actuators, Volume II

Weidong Wang [1,*], **Yong Ruan** [2], **Zaifa Zhou** [3] **and Min Liu** [1]

[1]   School of Mechano-Electronic Engineering, Xidian University, Xi'an 710071, China; mliu_12@stu.xidian.edu.cn
[2]   Department of Precision Instruments, Tsinghua University, Beijing 100084, China; ruanyong@mail.tsinghua.edu.cn
[3]   School of Electronic Science and Technology, Southeast University, Nanjing 210096, China; zfzhou@seu.edu.cn
*   Correspondence: wangwd@mail.xidian.edu.cn

Citation: Wang, W.; Ruan, Y.; Zhou, Z.; Liu, M. Design and Fabrication of Micro/Nano Sensors and Actuators, Volume II. *Micromachines* **2024**, *15*, 667.
https://doi.org/10.3390/mi15060667

Received: 13 May 2024
Accepted: 18 May 2024
Published: 21 May 2024

Microelectromechanical system (MEMS) sensors are a miniaturized sensor technology that integrates sensors with microelectronic components using microelectromechanical system manufacturing technology [1,2]. MEMS sensors typically consist of micromechanical structures, sensing circuits, and signal processing circuits, which can be used to measure and detect various physical quantities, such as pressure [3,4], temperature [5,6], acceleration [7,8], angular velocity [9,10], and humidity [11,12]. These micromechanical structures are typically composed of microsprings [13], thin films [14], and cantilever beams [15], and their dimensions typically range from micrometers to millimeters. When external physical quantities act on these micromechanical structures, they will undergo deformation or vibration, thereby changing parameters such as current, voltage, or capacitance in the circuit and generating corresponding electrical signals [16–18]. The miniaturization and integration of MEMS sensors have made them applicable in various fields, such as the automotive industry [19], healthcare [20], and aerospace [21], and have provided important support for the development of intelligent systems [22] and the Internet of Things [23].

This Special Issue contains 10 papers, mainly introducing various MEMS sensors, including inertial devices such as accelerometers and gyroscopes (Contributions 1 and 2), differential pressure sensors (Contribution 3), optical analysis of microfluidic technology (Contribution 4), thermoelectric temperature sensors (Contribution 5), thin film thermal flow sensors (Contribution 6), thermocouple sensors (Contribution 7), resonators (Contribution 8), microwave sensors (Contribution 9), and skin friction sensors (Contribution 10).

In particular, Zhang et al. (Contribution 1) presented a MEMS electrochemical angular accelerometer with a silicon-based four-electrode structure. Numerical simulations were conducted to optimize the geometrical parameters of the silicon-based four-electrode structure. These simulations demonstrated that increases in fluid resistance and cathode area can expand the working bandwidth and improve the device sensitivity, respectively. Zhang et al. (Contribution 2) proposed a weak capacitance detection circuit for a micro-hemispherical gyroscope based on common-mode feedback fusion modulation and demodulation. The input common-mode feedback is applied to the circuit to address the input common-mode voltage drift caused by both parasitic capacitance and gain capacitance. Secondly, a low-noise, high-gain amplifier is used to reduce the equivalent input noise. Thirdly, the modulator–demodulator and filter are introduced to the proposed circuit to effectively mitigate the side effects of noise. Yasunaga et al. (Contribution 3) designed a MEMS differential pressure sensor with a dynamic pressure canceler. They proposed a sensor "cap" that cancels the wind effect and noise by utilizing the airflow around a sphere. In addition, a height estimation method was developed based on a discrete transfer function model. This sensor can be used for outdoor drones and wave height measurement. Qin et al. (Contribution 4) analyzed the optical path in non-uniform refractive index media in surface acoustic wave (SAW) microfluidic chips and proposed a MEMS SAW device based on a solid medium. The device can control the focal length by changing the voltage and adjusting the sharpness of the micrograph. Wang et al. (Contribution 5) proposed a flexible temperature sensor based on graphene fibers. Graphene fibers (GFs) are synthesized by a facile microfluidic spinning technique using a green reducing agent (vitamin C). The sensor has good scalability in length, and its sensitivity can increase with the number of p–n thermocouples. Chen et al. (Contribution 6) established a finite element analysis model of thin-film heat flow meters based on finite element simulation methods, and a double-type thin-film heat flow sensor based on a copper/concentrate thermopile was made. Moreover, the device exhibited a dual-use characteristic, which considerably broadened the scope of its applications and service life and introduced a novel approach to measuring heat flow density in the intricate environment on the surface of the aeroengine. Han et al. (Contribution 7) studied the impact of different annealing times on the cyclic characteristics of ceramic oxide thin film thermocouples. The results show that when the annealing temperature is held constant, extending the annealing time can improve the properties of the film, increase its density, slow down oxidation, and enhance the thermal stability of the thermocouple. Liu et al. (Contribution 8) conducted in-depth analysis on the design and optimization of hemispherical resonators. The geometric parameters that significantly contribute to the performance of the resonator were determined via a thermoelastic model and process characteristics. And then they proposed an optimization method for the MEMS polysilicon hemispherical resonator based on PSO-BP and NSGA-II. Kim et al. (Contribution 9) designed a multi-metal crack detection (MCD) microwave sensor with two independent complementary split-ring resonators (CSRRs) loaded onto the substrate-integrated waveguide (SIW). The concentrated electromagnetic field of the SIW improves the Q-factor of the CSRRs. The proposed multi-MCD sensor can detect cracks of different widths and depths with a high Q-factor and high precision. Guo et al. (Contribution 10) studied the three-dimensional numerical simulation and structural optimization for a MEMS skin friction sensor in hypersonic flow. The objective of this study is to investigate the influence of sensor-sensitive structure on wall-flow characteristics and friction measurement accuracy. To this end, the two-dimensional and three-dimensional numerical models of the sensor in the hypersonic flow field based on computational fluid dynamics (CFD) are presented. Finally, the sensor-sensitive unit structure's design criterion is obtained to improve skin friction's measurement accuracy.

We would like to take this opportunity to thank all the authors who have submitted their papers to this Special Issue and to thank all the reviewers for their contributions in evaluating the submitted papers and for their efforts in improving the quality of the manuscripts.

Conflicts of Interest: The authors declare no conflicts of interest.

List of Contributions:

1. Zhang, M.; Liu, Q.; Zhu, M.; Chen, J.; Chen, D.; Wang, J.; Lu, Y. A MEMS Electrochemical Angular Accelerometer with Silicon-Based Four-Electrode Structure. *Micromachines* **2024**, *15*, 351. https://doi.org/10.3390/mi15030351
2. Zhang, X.; Li, P.; Zhuang, X.; Sheng, Y.; Liu, J.; Gao, Z.; Yu, Z. Weak Capacitance Detection Circuit of Micro-Hemispherical Gyroscope Based on Common-Mode Feedback Fusion Modulation and Demodulation. *Micromachines* **2023**, *14*, 1161. https://doi.org/10.3390/mi14061161
3. Yasunaga, S.; Takahashi, H.; Takahata, T.; Shimoyama, I. MEMS Differential Pressure Sensor with Dynamic Pressure Canceler for Precision Altitude Estimation. *Micromachines* **2023**, *14*, 1941. https://doi.org/10.3390/mi14101941
4. Qin, X.; Chen, X.; Yang, Q.; Yang, L.; Liu, Y.; Zhang, C.; Wei, X.; Wang, W. Analysis of Acousto-Optic Phenomenon in SAW Acoustofluidic Chip and Its Application in Light Refocusing. *Micromachines* **2023**, *14*, 943. https://doi.org/10.3390/mi14050943
5. Wang, C.; Zhang, Y.; Han, F.; Jiang, Z. Flexible Thermoelectric Type Temperature Sensors Based on Graphene Fibers. *Micromachines* **2023**, *14*, 1853. https://doi.org/10.3390/mi14101853

6. Chen, H.; Liu, T.; Feng, N.; Shi, Y.; Zhou, Z.; Dai, B. Structural Design of Dual-Type Thin-Film Thermopiles and Their Heat Flow Sensitivity Performance. *Micromachines* **2023**, *14*, 1458. https://doi.org/10.3390/mi14071458
7. Han, Y.; Ruan, Y.; Xue, M.; Wu, Y.; Shi, M.; Song, Z.; Zhou, Y.; Teng, J. Effect of Annealing Time on the Cyclic Characteristics of Ceramic Oxide Thin Film Thermocouples. *Micromachines* **2022**, *13*, 1970. https://doi.org/10.3390/mi13111970
8. Liu, J.; Li, P.; Zhuang, X.; Sheng, Y.; Qiao, Q.; Lv, M.; Gao, Z.; Liao, J. Design and Optimization of Hemispherical Resonators Based on PSO-BP and NSGA-II. *Micromachines* **2023**, *14*, 1054. https://doi.org/10.3390/mi14051054
9. Kim, Y.; Park, E.; Salim, A.; Kim, J.; Lim, S. Microwave DualCrack Sensor with a High Q-Factor Using the TE20 Resonance of a Complementary Split-Ring Resonator on a Substrate-Integrated Waveguide. *Micromachines* **2023**, *14*, 578. https://doi.org/10.3390/mi14030578
10. Guo, H.; Wang, X.; Liu, T.; Guo, Z.; Gao, Y. 3D Numerical Simulation and Structural Optimization for a MEMS Skin Friction Sensor in Hypersonic Flow. *Micromachines* **2022**, *13*, 1487. https://doi.org/10.3390/mi13091487

References

1. de Groot, W.A.; Webster, J.R.; Felnhofer, D.; Gusev, E.P. Review of device and reliability physics of dielectrics in electrostatically driven MEMS devices. *IEEE Trans. Device Mater. Reliab.* **2009**, *9*, 190–202. [CrossRef]
2. Zhu, J.; Liu, X.; Shi, Q.; He, T.; Sun, Z.; Guo, X.; Lee, C. Development trends and perspectives of future sensors and MEMS/NEMS. *Micromachines* **2019**, *11*, 7. [CrossRef] [PubMed]
3. Liu, Y.; Jiang, X.; Yang, H.; Qin, H.; Wang, W. Structural Engineering in Piezoresistive Micropressure Sensors: A Focused Review. *Micromachines* **2023**, *14*, 1507. [CrossRef] [PubMed]
4. Nallathambi, A.; Shanmuganantham, T.; Sindhanaiselvi, D. Design and analysis of MEMS based piezoresistive pressure sensor for sensitivity enhancement. *Mater. Today Proc.* **2018**, *5*, 1897–1903. [CrossRef]
5. Kose, T.; Azgin, K.; Akin, T. Design and fabrication of a high performance resonant MEMS temperature sensor. *J. Micromech. Microeng.* **2016**, *26*, 045012. [CrossRef]
6. Algamili, A.S.; Khir, M.H.; Ahmed, A.Y.; Rabih, A.A.; Ba-Hashwan, S.S.; Alabsi, S.S.; Al-Mahdi, O.L.; Isyaku, U.B.; Ahmed, M.G.; Junaid, M. Fabrication and Characterization of the Micro-Heater and Temperature Sensor for PolyMUMPs-Based MEMS Gas Sensor. *Micromachines* **2022**, *13*, 525. [CrossRef] [PubMed]
7. Liu, M.; Wu, X.; Niu, Y.; Yang, H.; Zhu, Y.; Wang, W. Research Progress of MEMS Inertial Switches. *Micromachines* **2022**, *13*, 359. [CrossRef] [PubMed]
8. Singh, V.; Kumar, V.; Saini, A.; Khosla, P.K.; Mishra, S. Design and development of the MEMS-based high-g acceleration threshold switch. *J. Microelectromech. Syst.* **2020**, *30*, 24–31. [CrossRef]
9. Ren, J.; Zhou, T.; Zhou, Y.; Li, Y.; Su, Y. An In-Run Automatic Mode-Matching Method with Amplitude Correction and Phase Compensation for MEMS Disk Resonator Gyroscope. *IEEE Trans. Instrum. Meas.* **2023**, *72*, 7505911. [CrossRef]
10. Li, Z.; Cui, Y.; Gu, Y.; Wang, G.; Yang, J.; Chen, K.; Cao, H. Temperature drift compensation for four-mass vibration MEMS gyroscope based on EMD and hybrid filtering fusion method. *Micromachines* **2023**, *14*, 971. [CrossRef]
11. Dennis, J.O.; Ahmed, A.Y.; Khir, M.H. Fabrication and characterization of a CMOS-MEMS humidity sensor. *Sensors* **2015**, *15*, 16674–16687. [CrossRef] [PubMed]
12. Hassan, A.S.; Juliet, V.; Raj, C.J.A. MEMS bas ed humidity sensor with integration of temperature sensor. *Mater. Today Proc.* **2018**, *5*, 10728–10737. [CrossRef]
13. Zhang, F.; Wang, C.; Yuan, M.; Tang, B.; Xiong, Z. Conception, fabrication and characterization of a silicon based MEMS inertial switch with a threshold value of 5 g. *J. Micromech. Microeng.* **2017**, *27*, 125001. [CrossRef]
14. Song, X.; Liu, H.; Fang, Y.; Zhao, C.; Qu, Z.; Wang, Q.; Tu, L.C. An integrated gold-film temperature sensor for in situ temperature measurement of a high-precision MEMS accelerometer. *Sensors* **2020**, *20*, 3652. [CrossRef]
15. Xu, Q.; Rocha, R.T.; Algoos, Y.; Feron, E.; Younis, M.I. Design, simulation, and testing of a tunable MEMS multi-threshold inertial switch. *Microsyst. Nanoeng.* **2024**, *10*, 31. [CrossRef] [PubMed]
16. Zhang, J.; Chen, J.; Li, M.; Ge, Y.; Wang, T.; Shan, P.; Mao, X. Design, fabrication, and implementation of an array-type MEMS piezoresistive intelligent pressure sensor system. *Micromachines* **2018**, *9*, 104. [CrossRef] [PubMed]
17. Zhao, Y.; Wang, Z.; Tan, S.; Liu, Y.; Chen, S.; Li, Y.; Hao, Q. Dependance of Gauge Factor on Micro-Morphology of Sensitive Grids in Resistive Strain Gauges. *Micromachines* **2022**, *13*, 280. [CrossRef]
18. Cao, Y.; Xi, Z.; Yu, P.; Wang, J.; Nie, W. A MEMS inertial switch with a single circular mass for universal sensitivity. *J. Micromech. Microeng.* **2015**, *25*, 105005. [CrossRef]
19. Zhang, W.; Hao, C.; Zhang, Z.; Yang, S.; Peng, J.; Wu, B.; Wang, R. Vector high-resolution marine turbulence sensor based on a MEMS bionic cilium-shaped structure. *IEEE Sens. J.* **2020**, *21*, 8741–8750. [CrossRef]
20. Baik, J.; Seo, S.; Lee, S.; Yang, S.; Park, S.M. Circular radio-frequency electrode with MEMS temperature sensors for laparoscopic renal sympathetic denervation. *IEEE Trans. Biomed. Eng.* **2021**, *69*, 256–264. [CrossRef]

21. Ghazali, M.H.M.; Rahiman, W. Fuzzy-Oriented Anomaly Inspection in Unmanned Aerial Vehicle (UAV) Based on MEMS Accelerometers in Multi-Mode Environment. *IEEE Trans. Instrum. Meas.* **2023**, *72*, 3530710. [CrossRef]
22. Podder, I.; Fischl, T.; Bub, U. Artificial Intelligence Applications for MEMS-Based Sensors and Manufacturing Process Optimization. *Telecom* **2023**, *4*, 165–197. [CrossRef]
23. Xu, Q.; Yang, Z.; Sun, Y.; Lai, L.; Jin, Z.; Ding, G.; Wang, J. Shock-resistibility of mems-base d inertial microswitch under reverse directional ultra-high g acceleration for IoT applications. *Sci. Rep.* **2017**, *7*, 45512. [CrossRef] [PubMed]

micromachines

Article

A MEMS Electrochemical Angular Accelerometer with Silicon-Based Four-Electrode Structure

Mingbo Zhang [1,2], Qinghua Liu [1,2], Maoqi Zhu [1,2], Jian Chen [1,2], Deyong Chen [1,2], Junbo Wang [1,2] and Yulan Lu [1,2,*]

1   State Key Laboratory of Transducer Technology, Aerospace Information Research Institute, Chinese Academy of Sciences, Beijing 100190, China; zhangmingbo21@mails.ucas.ac.cn (M.Z.); liuqinghua21@mails.ucas.ac.cn (Q.L.); zhumaoqi21@mails.ucas.ac.cn (M.Z.); chenjian@mail.ie.ac.cn (J.C.); dychen@mail.ie.ac.cn (D.C.); jbwang@mail.ie.ac.cn (J.W.)

2   School of Electronic, Electrical and Communication Engineering, University of Chinese Academy of Sciences, Beijing 100049, China

*   Correspondence: luyl@aircas.ac.cn; Tel.: +86-010-58887184

Abstract: This paper presents a MEMS electrochemical angular accelerometer with a silicon-based four-electrode structure, which was made of thousands of interconnected microchannels for electrolyte flow, anodes uniformly coated on structure surfaces and cathodes located on the sidewalls of flow holes. From the perspective of device fabrication, in this study, the previously reported multi-piece assembly was simplified into single-piece integrative manufacturing, effectively addressing the problems of complex assembly and manual alignment. From the perspective of the sensitive structure, in this study, the silicon-based four-electrode structure featuring with complete insulation layers between anodes and cathodes can enable fast electrochemical reactions with improved sensitivities. Numerical simulations were conducted to optimize the geometrical parameters of the silicon-based four-electrode structure, where increases in fluid resistance and cathode area were found to expand working bandwidths and improve device sensitivity, respectively. Then, the silicon-based four-electrode structure was fabricated by conventional MEMS processes, mainly composed of wafer-level bonding and wafer-level etching. As to device characterization, the MEMS electrochemical angular accelerometer with the silicon-based four-electrode structure exhibited a maximum sensitivity of 1458 V/(rad/s^2) at 0.01 Hz and a minimum noise level of -164 dB at 1 Hz. Compared with previously reported electrochemical angular accelerometers, the angular accelerometer developed in this study offered higher sensitivities and lower noise levels, indicating strong potential for applications in the field of rotational seismology.

Keywords: microelectromechanical system; electrochemical angular accelerometer; rotational sensor; silicon-based four-electrode structure

Citation: Zhang, M.; Liu, Q.; Zhu, M.; Chen, J.; Chen, D.; Wang, J.; Lu, Y. A MEMS Electrochemical Angular Accelerometer with Silicon-Based Four-Electrode Structure. *Micromachines* **2024**, *15*, 351. https://doi.org/10.3390/mi15030351

Academic Editor: Ha Duong Ngo

Received: 9 February 2024
Revised: 24 February 2024
Accepted: 27 February 2024
Published: 29 February 2024

1. Introduction

Seismic motions consist of both translational and rotational components, resulting in vibrations with six degrees of freedom [1,2]. However, traditional seismic measurements only focus on translational components without paying attention to rotational aspects, which thus cannot fully meet the requirements for seismic resistance [3]. In fact, in recent years, a series of seismic damages caused by rotational vibrations have been recorded [4–7]. Due to the lack of effective measurements for angular accelerations, especially for angular accelerometers with low frequencies down to 0.01 Hz, the understanding of rotational components remains limited.

Angular accelerometers for seismic motion monitoring are sensors that convert low-frequency angular accelerations from external environments into electrical signals. The potential application is to monitor and prevent the hazards caused by the rotational components of seismic motions and to comprehensively analyze the causes of earthquakes.

Current angular accelerometers for measuring seismic rotational components can be classified into three types based on solid inertial masses, sensitive microstructures and liquid inertial masses, respectively. Angular accelerometers based on solid inertial masses [8–10] had limited low-frequency performances and bulky sizes due to their wheel servo systems. Angular accelerometers based on sensitive microstructures [11–13] utilized sensitive principles of piezoelectric or piezoresistive effects enabled by microstructures (e.g., microcolumns or microbeams), and they were susceptible to linear vibrations and had limited frequency ranges and accuracies. Angular acceleration sensors based on liquid inertial masses [14–16] operated in a low-pass manner with high electromechanical conversion coefficients, demonstrating strong potential for detecting seismic motions with low amplitudes at low and ultra-low frequencies. The sensors working under the mechanism of electrochemical reaction have the merits of good adaptability [2] and controllability, and they are also highly sensitive to the ion concentration, thus achieving high sensitivity.

Traditional angular accelerations based on liquid inertial masses (electrochemical angular accelerometers) relied on platinum mesh electrodes as sensitive units [14], which suffered from the key problem of low consistency due to the misalignments among multiple layers of electrodes. Furthermore, the insulation layers of platinum meshes were made of porous ceramics, in which the production process was complicated and parameter adjustment accuracy was limited, further limiting measurement sensitivities. Therefore, optimizing the structure design and fabrication process of the electrode chip is key to the development of an electrochemical angular accelerometer. Since MEMS offered advantages such as miniaturization [2], integration, and mass production [17], a few pioneering studies were conducted to replace traditional platinum mesh electrodes with microfabricated electrodes, where both the sensitivities and working bandwidth of electrochemical angular accelerometers were significantly improved. However, these previously developed angular accelerometers with microfabricated micro-electrodes suffered from electrode structure and inter-electrode wires because of the planar setup of sensitive microelectrodes. The previously reported MEMS electrochemical angular accelerometers mostly comprise a silicon-based three-electrode structure [15] and a planar electrode structure [16]. Among them, the ones with a silicon-based three-electrode structure have higher sensitivity. But the chip did not have an active ion depletion layer, which would lead to unstable differential output. The parameters of an MEMS electrochemical angular accelerometer based on a planar electrode could be precisely adjusted. However, the small cathode area and large substrate area result in a low utilization efficiency of the on-chip area, limiting the sensitivity level within the frequency band.

This paper developed a MEMS electrochemical angular accelerometer to monitor the rotational components of seismic motions. A unique silicon-based four-electrode structure was included, which was made of thousands of interconnected microchannels for electrolyte flow, anodes uniformly coated on wafer surfaces, and cathodes located on the sidewalls of flow holes. From the perspective of device fabrication, the previously reported multi-piece assembly was simplified into single-piece integrative manufacture. The chip innovatively adopted a wafer-level bonding method based on SU-8, effectively addressing the problems of complex assembly and manual alignment. SU-8 serving as an intermediate layer existed around each hole and acted as both an insulation layer between the top and bottom cathodes and an ion depletion layer, which enabled stable output. From the perspective of the sensitive structure, in this study, the silicon-based four-electrode structure featured complete insulation layers between anodes and cathodes. The cathodes located on the sidewalls of flow holes were connected to each other by conduction through the silicon substrate, eliminating complex leads and improving the utilization efficiency of the on-chip area, which can enable fast electrochemical reactions with improved sensitivities.

2. Structure and Working Principle

The MEMS electrochemical angular accelerometer based on the silicon-based four-electrode structure is illustrated in Figure 1a. It consists of sensitive electrodes, a toroid channel, and an electrolyte solution. The toroid channel is filled with the electrolyte solution composed of I_2 and KI. The sensitive electrodes are fixed inside the toroid channel, where two identical pairs of electrodes are arranged in an A-C-C-A (anode-cathode-cathode-anode) configuration, separated by insulation layers.

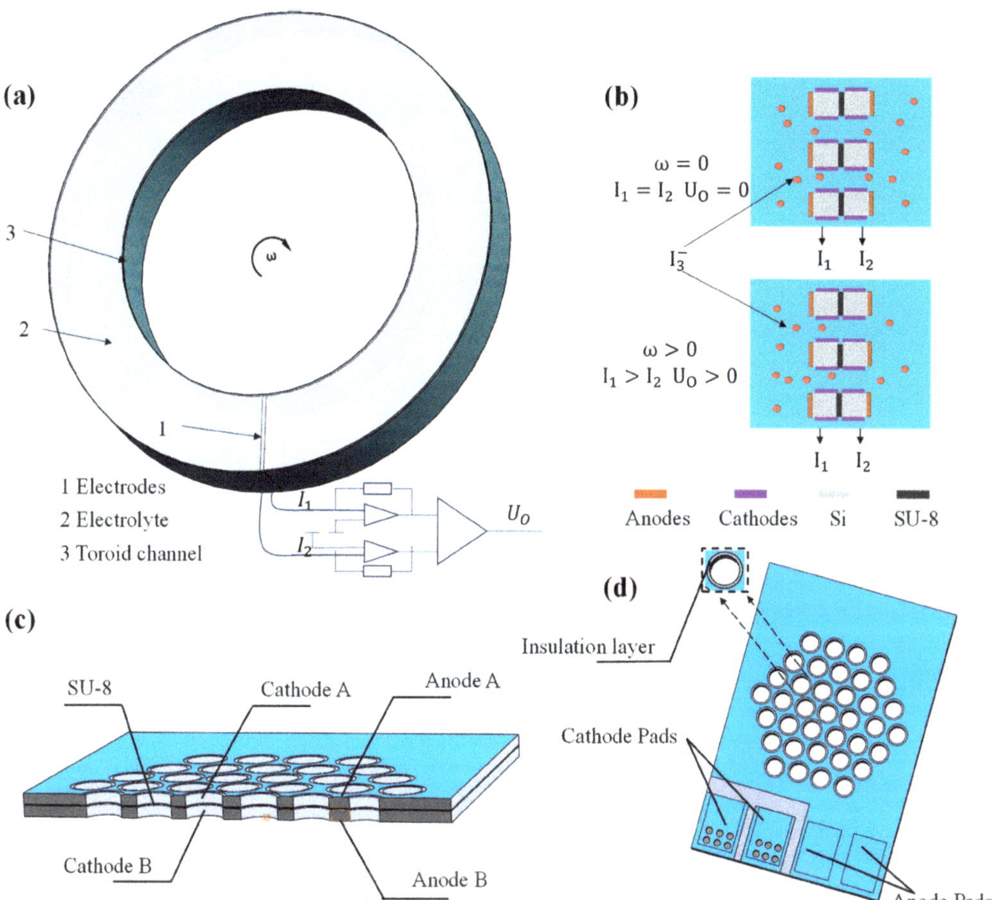

Figure 1. (**a**) The schematic of the MEMS electrochemical angular accelerometer with a silicon-based four-electrode structure, consisting of sensitive electrodes, a circular flow channel, and an electrolyte solution. (**b**) When the accelerometer detects external angular vibrations, there exists relative motions between the liquid inertial mass and the electrodes, where the I_3^- concentration near the two cathodes changes in magnitude but with opposite directions, leading to current outputs. (**c**) The cross-section profile, and (**d**) the 3D schematic view of the silicon-based four-electrode structure which is based on two silicon wafers containing thousands of through-hole microchannels for the electrolyte solution. The sidewalls of the flow holes and the surfaces of the bonded silicon wafers are uniformly sputtered with Ti/Pt alloy as cathodes and anodes, respectively.

Due to the presence of the complexation reaction $I_2 + I^- \rightleftharpoons I_3^-$, I^- and I_3^- in the electrolyte solution, these are the main components involved in the electrochemical reactions. During the operation of the electrochemical angular accelerometer, a high potential is applied to the surface-sensitive electrodes, which are regarded as two anodes, while a low potential is applied to the sensitive electrodes on the sidewalls of the flow holes, which function as two cathodes. When a voltage difference exists between anodes and cathodes, the following electrochemical reactions (anode: $3I^- - 2e^- \rightarrow I_3^-$, cathode: $I_3^- + 2e^- \rightarrow 3I^-$) occur.

The working principle of the MEMS electrochemical angular accelerometer is illustrated in Figure 1b. When the accelerometer does not detect any external angular vibrations, there is no relative motion between the electrolyte and the sensitive electrodes, resulting in a stable distribution of I_3^- near the sensitive electrodes. The output currents of the two cathodes are equal ($I_1 = I_2$), and the differential output signal is zero ($U_O = 0$). When the MEMS electrochemical angular accelerometer detects external angular vibrations, it induces relative motions between the inertial liquid mass and the electrodes. Through convection and diffusion effects, the concentrations of I_3^- near-two cathodes change in the same magnitudes but in opposite directions, causing the output current to change with direction ($I_1 > I_2$).

The cross-section profile of the silicon-based four-electrode structure is shown in Figure 1c, which is based on two bonded silicon wafers containing thousands of through-hole microchannels for the electrolyte solution. The sidewalls of the flow holes and the surfaces of the bonded silicon wafers are uniformly sputtered with Ti/Pt alloy as cathodes and anodes, respectively. More specifically, cathodes are electrically connected through the silicon substrate with resistance less than 0.0015 Ω/cm, and external connectivity is maintained by sputtering metal after etching flow holes. The anodes are directly connected to silicon surfaces, and anode-cathode insulation is realized by the gap between anodes and cathodes.

The three-dimensional schematic of the silicon-based four-electrode structure is shown in Figure 1d. Each side has two wire-bonding pads for the same electrode, with a total of eight wire-bonding pads on the top and bottom sides, resulting in four electrodes. The wire bonding pads are connected to the signal acquisition circuit through wire bonding, and the silicon-based four-electrode structure is encapsulated in a casing for testing. The bonding between the silicon wafers is completed during the wafer fabrication process, where SU-8 serves as an intermediate layer and acts as both an insulation layer between the top and bottom cathodes and an ion depletion layer.

3. Theory and Numerical Simulation

The relative motions between the electrolyte solution and the silicon-based four-electrode structure follow the principle of fluid mechanics, and the electrochemical reactions near the sensitive electrodes adhere to Faraday's law. Therefore, flow resistance and cathode area are key parameters that influence the performance of the MEMS electrochemical angular accelerometer [18,19]. Since the numbers and diameters of flow holes and cathode length determine the flow resistance and cathode area of this angular accelerometer, they are treated as crucial geometrical parameters in the design and optimization of the silicon-based four-electrode structure.

Since theoretical analysis can only provide a qualitative rather than quantitative relationship between the aforementioned geometrical parameters and the performance of the MEMS electrochemical angular accelerometer, numerical simulations were conducted in this study for structural optimization based on numerical results.

In COMSOL, the simulation model included the sensitive unit, its adjacent flow channels, and the internal electrolyte. The equations in fluid mechanics and Faraday's law were applied, and the properties of incompressible fluid, ion transports and electrode boundary conditions were properly defined. Based on these settings, a two-dimensional simulation was established based on the transverse cross section near the electrode of the electrochemical angular accelerometer, as shown in Figure 2a.

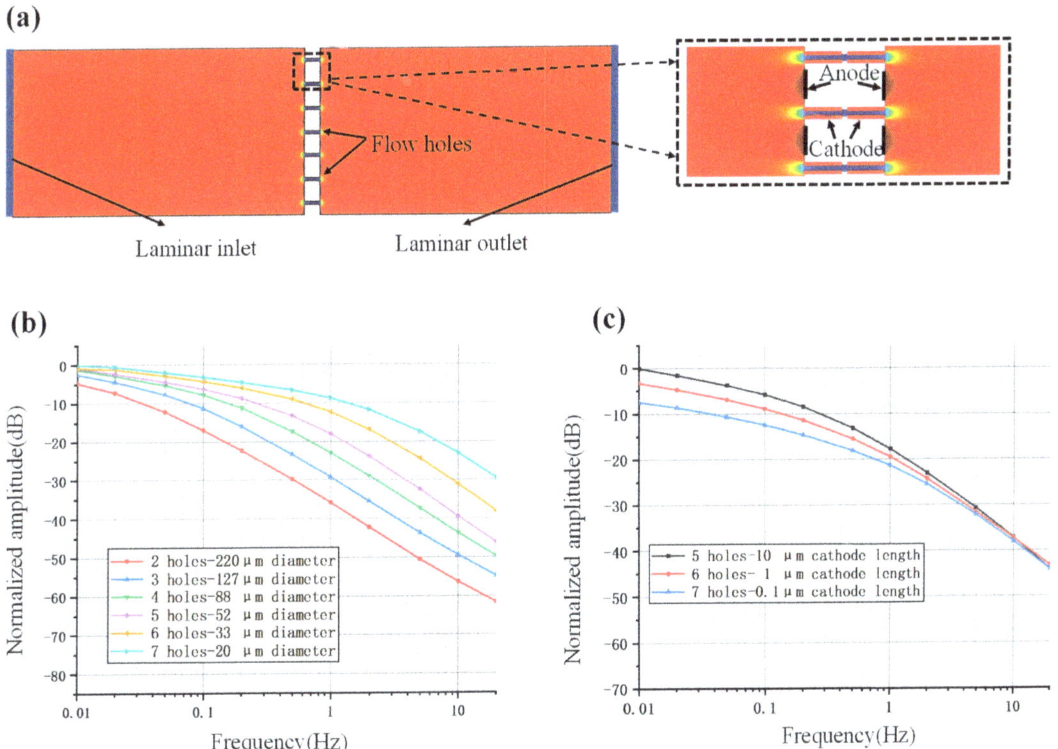

Figure 2. Numerical Simulation of the MEMS electrochemical angular accelerometer with silicon-based four-electrode structure. (**a**) A two-dimensional simulation is established based on the transverse cross section near the electrodes of the electrochemical angular accelerometer, where the simulation model includes the sensitive unit, its adjacent flow channels, and electrode boundary conditions. Simulation results for the silicon-based four-electrode structure with the influence of (**b**) different fluid resistance on the bandwidth and (**c**) different cathode area on the sensitivity where increases in fluid resistance and cathode area were found to expand working bandwidths and improve device sensitivity, respectively.

Figure 2b shows simulation results for the influence of different fluid resistances on the bandwidth of the silicon-based four-electrode structure. The parameters used for simulation are shown in Table 1. More specifically, under a unit length with a cross section of 500 μm, a variety of number-diameter parameters of flow holes (e.g., 2 hole-220 μm of diameter, 3 hole-127 μm of diameter, 4 hole-88 μm of diameter, 5 hole-52 μm of diameter, 6 hole-33 μm of diameter, 7 hole-20 μm of diameter) were simulated for comparison under the unchanged cathode length of 100 μm for each hole. As to the simulation results, it was found that increases in flow resistance produced accelerometers with increased working bandwidth due to the turning point of the low-pass link shifting to the right, and the attenuation of the mid- to high-frequency curve slowing down.

Table 1. Normalized flow resistance corresponding to different parameters.

Length of Cross Section	Number of Holes	Diameter of Holes	Normalized Flow Resistance
500 μm	2	220	0.000239
500 μm	3	127	0.001435
500 μm	4	88	0.04669
500 μm	5	52	0.030636
500 μm	6	33	0.157402
500 μm	7	20	1

Figure 2c shows the simulation results for the influence of different cathode areas on the sensitivity of the silicon-based four-electrode structure. More specifically, under a unit lengths with a cross section of 500 μm, under predefined number-diameter parameters of flow holes (e.g., 5 hole-52 μm of diameter, 6 hole-33 μm of diameter, 7 hole-20 μm of diameter), cathode length of 10 μm, 1 μm and 0.1 μm were simulated for comparison. As to the simulation results, it was found that increases in cathode length produced accelerometers with increased sensitivities due to increased rates of electrochemical reactions.

The simulation provides a basis for determining the effects of two key parameters, fluid resistance and cathode area, on device performance, which is crucial for chip parameter design and device selection. However, since two-dimensional simulations do not align perfectly with actual characteristics, devices with different parameters will be tested during actual fabrication. Parameters will be optimized and analyzed based on the conclusions drawn from the simulations.

4. Fabrication and Packaging

The silicon-based four-electrode structure was fabricated by conventional MEMS processes, which were divided mainly into two steps of wafer-level bonding and wafer-level etching. In the step of wafer-level bonding, SU-8 was used as the adhesive layer because of its high insulation and stability to bond two silicon wafers with a thickness of 200 μm and a diameter of 4 inches. The wafer-level bonding based on SU-8 is shown in Figure 3a, which was composed of (i) cleaning and oxidation of silicon wafers with a thickness of 1 μm based on plasma enhanced chemical vapor deposition (PECVD), (ii) Spin-coating of SU-8 photoresist on silicon wafers, and (iii) bonding with alignment enabled by a laminating machine for compression and a hot plate for drying.

The bonded wafers were manufactured into the silicon-based four-electrode structures following the MEMS process shown in Figure 3b, which was described as follows.

Step 1. Washing the bonded silicon wafer by boiling acid and deionized water.

Step 2. Spin-coating AZ1500 on both sides of the bonded silicon wafer, followed by prebaking, lithography, and development to function as a mask for anode sputtering.

Step 3. Sputtering Ti/Pt alloy with a thickness of 300/2500 Å on both sides of the silicon wafer, followed by lift-off to obtain anodes.

Step 4. Spin-coating AZ4620 on the front side of the bonded silicon wafer, followed by prebaking, lithography, and development to function as a mask for front-side etching.

Step 5. RIE etching the front side to remove the 1 μm silicon oxide layer, and DRIE etching the 200 μm silicon substrate to form frontside flow-through holes.

Step 6. Spin-coating AZ4620 on the backside of the bonded silicon wafer, followed by prebaking, lithography, and development to function as a mask for backside etching.

Step 7. RIE etching the backside to remove a 1 μm silicon oxide layer, and DRIE etching the 200 μm silicon substrate to form backside flow-through holes.

Step 8. RIE etching the intermediate SU-8 bonding layer with the etching gas of $CF_4:O_2 = 1:4$ to create fully penetrating flow-through holes.

Step 9. Attaching SD230 photoresist on both sides of the bonded silicon wafer, followed by prebaking, lithography, and development to function as a mask for cathode sputtering.

Step 10. Sputtering Ti/Pt alloy with a thickness of 300/2500 Å on both sides, followed by lift-off to obtain cathodes.

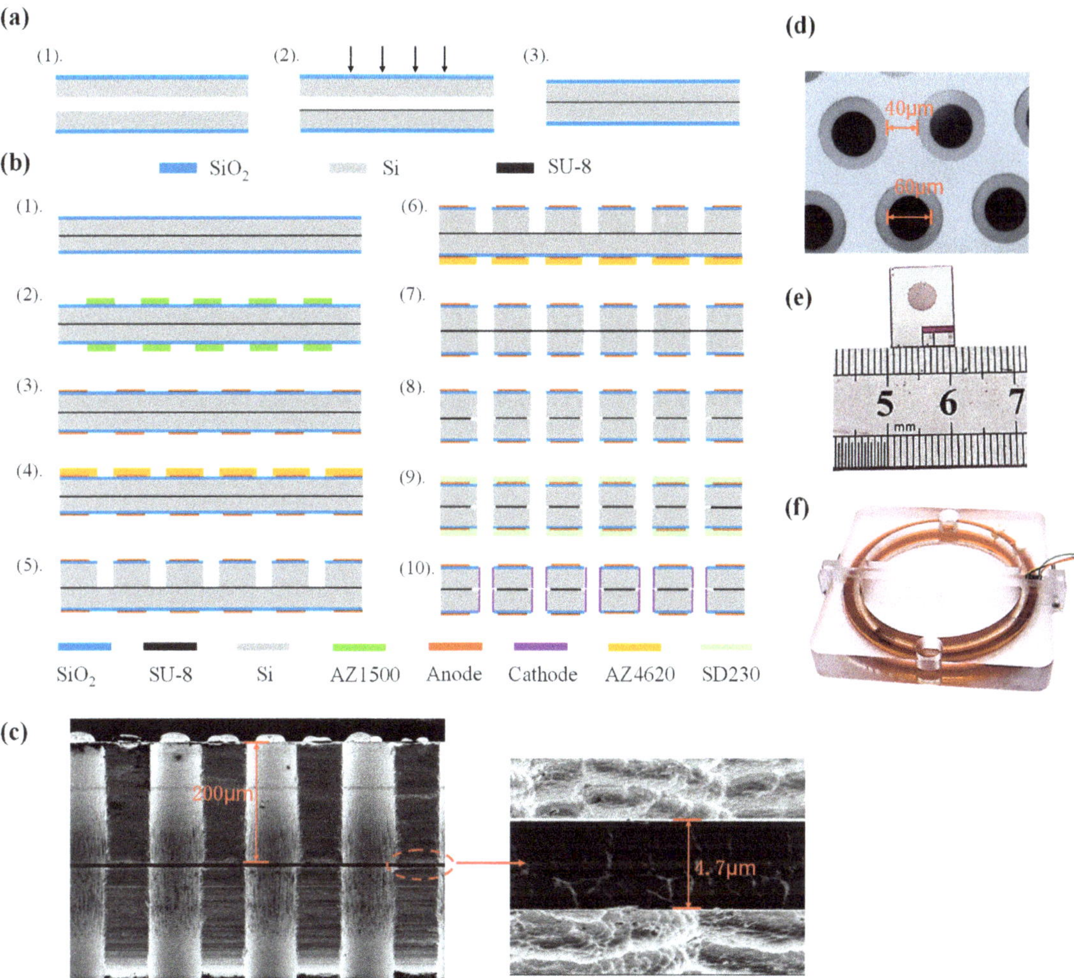

Figure 3. (**a**) Process of wafer-level bonding, including key steps of PECVD, Spin-coating of SU-8, compression and drying. (**b**) MEMS process for fabricating the silicon-based four-electrode structure, including sputtering of anodes, etching of flow channels, etching of SU-8, and sputtering of cathodes. (**c**) The cross-section profile of the flow-through holes by SEM, (**d**) The surface of the fabricated silicon-based four-electrode structure by electron microscope. (**e**) The image of the fabricated silicon-based four-electrode structure. (**f**) The assembled MEMS electrochemical angular accelerometer with the silicon-based four-electrode structure.

The cross-sectional profiles of the flow-through holes were captured using a scanning electron microscope (SEM), and the locally magnified image confirmed the uniformity of the intermediate layer SU-8, as shown in Figure 3c. It can be clearly seen that through the etching of RIE and DRIE, a through flow hole of 200 µm above and below was fabricated on the silicon-based four-electrode structure of the chip, and the thickness of the intermediate SU-8 layer is 4.7 µm. Figure 3d shows a microscopic view of the flow hole structure of the bonded silicon wafer before dicing. The black part is the flow holes with sidewall cathodes, and the gray part is the silicon dioxide insulation rings, while the white part is the anodes. The electrode structure shown in Figure 3d has a flow hole with a diameter of 60 μm and an anode width of 40 μm. The flow holes are consistent in size and evenly distributed. Figure 3e shows a chip of the silicon-based four-electrode structure after dicing, with a size of 10.4 mm × 13.7 mm. The middle circular part is densely distributed with thousands of flow holes, and cathodes on the sidewalls are led out to the lead hole (on the bottom right) through bulk silicon conduction. The fabricated silicon-based four-electrode structure was then assembled into an enclosure with mechanical compression, followed by the injection and sealing of the electrolyte solution (see Figure 3f). The dark part is an annular flow channel with a diameter of 120 mm. The channel was filled with electrolyte, and the chip was placed in the channel, which was the effective sensing part.

5. Results

The characterization system for the fabricated MEMS electrochemical angular accelerometer with the silicon-based integrated four-electrode structure included a standard angular vibration table that can provide sinusoidal vibrations, a DC power supply, a current-voltage conversion circuitry, a data acquisition card, and a LabVIEW-based data processing platform. The standard angular vibration table can provide a frequency signal ranging from 0.01 Hz to 10 Hz, and the corresponding output angular acceleration values at different frequencies can be adjusted by the excitation amplitudes.

The devices were assembled as follows: 1517 (number of flow holes)-60 (diameter of flow holes in μm)-160 (length of cathodes in μm), 1111-60-160, 1107-80-200, and 846-80-200. Note that the length of the insulation layer between anodes and cathodes was 20 μm. The assembled silicon-based integrated four-electrode structures of MEMS electrochemical angular accelerometers were placed on the angular vibration table, and a frequency signal ranging from 0.01 Hz to 10 Hz was provided. During the testing, it usually takes a period of time to stabilize the devices. And the stability of the device is determined by the consistency of multiple repeated tests. The obtained final sensitivity-frequency characteristics curve after stabilization is shown in Figure 4.

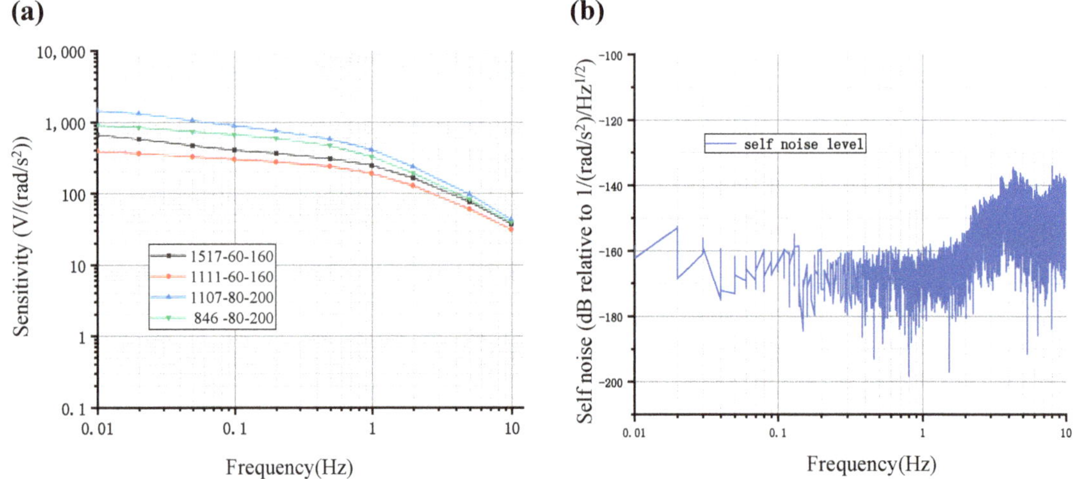

Figure 4. (**a**) Amplitude-frequency characteristic curves and (**b**) self-noise level for the MEMS electrochemical angular accelerometers with the silicon-based four-electrode structure under a group of geometrical parameters (1517-60-160, 1111-60-160, 1107-80-200 and 846-80-200).

The overall trend of the curve indicated a characteristic of a low-pass filter, where the sensitivity gradually decreased as the frequency was increased, which was consistent with the simulation results. Among them, the sensitivities of 1517-60-160 V.S. 1111-60-160 devices at frequencies of 0.01 Hz, 0.1 Hz, and 1 Hz were quantified as 665.7912 V.S. 390.2359 $V/(rad/s^2)$, 399.5451 V.S. 298.7483 $V/(rad/s^2)$, and 243.9414 V.S. 190.3566 $V/(rad/s^2)$, where the differences between these two frequency response curves were stable, respectively corresponding to the black and red lines in Figure 4a. Compared to the 1517-60-160 device and the 1111-60-160 device, they only differed in the number of holes. The 1517-60-160 device had a larger effective area of cathodes and a lower flow resistance due to a larger number of holes, and therefore had higher sensitivity and narrower bandwidth, while the 1111-60-160 device was the opposite. The sensitivities of the 1107-80-200 device vs. the 846-80-200 device at frequencies of 0.01 Hz, 0.1 Hz, and 1 Hz were quantified as 1458.9611 V.S. 905.3151 $V/(rad/s^2)$, 897.1403 V.S. 671.8375 $V/(rad/s^2)$, and 407.7017 V.S. 327.0718 $V/(rad/s^2)$, respectively, corresponding to the blue and green lines in Figure 4a. From the perspective of the number of holes, the 1107-80-200 device had a larger effective area of cathodes and a lower flow resistance than the 846-80-200 device.

Taking into account the bandwidth and sensitivity of the four devices, the 1107-80-200 device had the highest low-frequency sensitivity due to the largest effective cathode area while maintaining a wide frequency band. As the 1107-80-200 device had the best performance among the four devices, subsequent analysis would focus on the 1107-80-200 device.

The 1107-80-200 MEMS electrochemical angular accelerometer with the silicon-based integrated four-electrode structure was then placed on a flat surface and tested for self-noise level at night, where the self-noise power spectral density is shown in Figure 4b. The noise levels at 0.01 Hz, 0.1 Hz, 1 Hz, and 2 Hz were quantified as -161 dB, -164 dB, -162 dB, and -158 dB, respectively. The device exhibited stable low noise levels within the range of 0.01–2 Hz, with slightly increased noise values (lower than -140 dB) in the 2–10 Hz frequency range. Besides the high performance from the perspective of device sensitivity, the MEMS electrochemical angular accelerometer developed in this study also exhibited excellent signal-to-noise ratios.

The 1107-80-200 MEMS-based electrochemical angular accelerometer was subjected to artificial excitations along with the commercial device METR-03 for seismic monitoring. The test results, as shown in Figure 5a, indicated that both devices clearly captured multiple artificial excitations with consistent response times, where the correlation coefficient between the two devices was 0.9527. The signal amplitude outputted by the device's silicon-based integrated four-electrode structure was significantly larger than that of the METR-03 device. More specifically, after an artificial excitation at 11.9 s, the maximum amplitude from the device with the silicon-based integrated four-electrode structure was 0.86 V, while the METR-03 device only outputted 0.06 V. This comparison can be observed clearly in the local vibration signal amplification graph, as shown in Figure 5b.

Finally, the performance of the MEMS electrochemical angular accelerometer with the silicon-based four-electrode structure was compared to the previously reported MEMS electrochemical angular accelerometers with a silicon-based three-electrode structure [15], a planar electrode structure [16], and electrochemical angular accelerometers with platinum mesh electrodes [20], as shown in Table 2. Compared to the other three devices, this device had a very high sensitivity of 42 $V/(rad/s^2)$ at 10 Hz and the lowest noise level of -164 dB. The excellent performance of the silicon-based four-electrode structure is due to the complete and controllable SU-8 intermediate layer, stable differential output, and high on-chip electrode area utilization. The MEMS electrochemical angular accelerometer with the silicon-based four-electrode structure exhibited excellent electrode structures, efficient electrochemical reactions, and a well-developed device packaging process, resulting in noticeable low noise levels and high sensitivities.

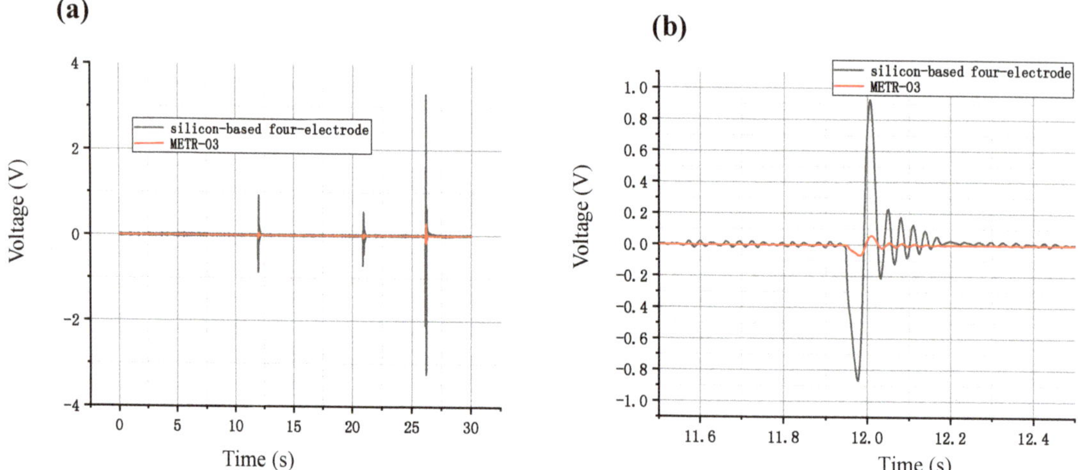

Figure 5. (**a**) Overall and (**b**) local voltage-time curves of the MEMS electrochemical angular accelerometer with the silicon-based four-electrode structure vs. the commercial counterpart of METR-03 in response to artificial vibrations.

Table 2. Comparison of key parameters of electrochemical angular accelerometers.

Characteristic	Unit	[20]	[15]	[16]	**This Device**
sensitivity	$V/(rad/s^2)$	1 @10 Hz	2.405 @10 Hz	0.033 @10 Hz	42 @10 Hz
noise level	$(rad/s^2)/Hz^{1/2}$	-105 dB	-154 dB	-118 dB	-164 dB

6. Conclusions

In this study, we present a MEMS electrochemical angular accelerometer with a silicon-based four-electrode structure that is made of thousands of interconnected microchannels for electrolyte flow, anodes uniformly coated on structure surfaces, and cathodes located on the sidewalls of flow holes. The silicon-based four-electrode structure was fabricated by conventional MEMS processes, mainly wafer-level bonding and wafer-level etching. The optimal parameters for the chip were ultimately determined as follows: the number of flow holes is 1107, the diameter of the flow hole is 80 μm, and the length of the cathode is 200 μm. As to device characterization, the MEMS electrochemical angular accelerometer with the silicon-based four-electrode structure exhibited a maximum sensitivity of 1458 $V/(rad/s^2)$ at 0.01 Hz and a minimum noise level of -164 dB at 1 Hz. The excellent performance of the silicon-based four-electrode structure is due to the complete and controllable SU-8 intermediate layer, stable differential output and high on-chip electrode area utilization. Compared with previously reported electrochemical angular accelerometers with platinum mesh electrodes and silicon-based three-electrode structures, the angular accelerometer developed in this study offered higher sensitivities and lower noise levels, indicating strong potential for applications to monitor and prevent the hazards caused by the rotational components of seismic motions and to comprehensively analyze the causes of earthquakes.

Author Contributions: Conceptualization, M.Z. (Mingbo Zhang), J.W. and D.C.; methodology, M.Z. (Mingbo Zhang); software, M.Z. (Mingbo Zhang), M.Z. (Maoqi Zhu) and Q.L.; validation, M.Z. (Mingbo Zhang); formal analysis, M.Z. (Mingbo Zhang); investigation, M.Z. (Mingbo Zhang); resources, M.Z. (Mingbo Zhang); data curation, M.Z. (Mingbo Zhang); writing—original draft preparation, M.Z. (Mingbo Zhang); writing—review and editing, J.W., D.C., J.C. and Y.L.; visualization, M.Z. (Mingbo Zhang); supervision, J.W., D.C., J.C. and Y.L.; project administration, J.W., D.C. and J.C.;

funding acquisition, J.W., D.C. and J.C. All authors have read and agreed to the published version of the manuscript.

Funding: This work was supported by the National Natural Science Foundation of China (52335012, 62201549, 62071454, 62061136012) and Beijing Natural Science Foundation (No. 4242012).

Data Availability Statement: Data are contained within the article.

Conflicts of Interest: The authors declare no conflicts of interests.

References

1. Hong, Z.; Cui, T. Study on the rotational component of earthquake ground motion. *Shanxi Archit.* **2012**, *38*, 39–40.
2. Hou, Y.; Jiao, R.; Yu, H. MEMS based geophones and seismometers. *Sens. Actuators A Phys.* **2021**, *318*, 112498. [CrossRef]
3. Jaroszewicz, L.; Kurzych, A.; Krajewski, Z.; Marć, P.; Kowalski, J.K.; Bobra, P.; Jankowski, R. Review of the usefulness of various rotational seismometers with laboratory results of fibre-optic ones tested for engineering applications. *Sensors* **2016**, *16*, 2161. [CrossRef] [PubMed]
4. Zhou, C.; Zeng, X.; Wang, Q.; Liu, W.; Wang, C. Construction of seismic rotational motion field for the Jiuzhaigou Ms7.0 earthquake based on ground tilt data. *Chin. Sci. Earth Sci.* **2019**, *49*, 811–821.
5. Wang, Y.; Tan, D.; Qiu, D.; Zhang, F.; Cheng, L. Path differences in the seismic waveform and Tianshidian electric field anomalies for the 2020 Yutian M_S6.4 earthquake in Xinjiang. *Earthquake* **2021**, *41*, 180–189.
6. Cao, Y.; Zeng, X.; Li, Z.; Wang, S.; Bao, F.; Xie, J. Analysis of rotational and translational components of surface wave records for the Yunnan Yangbi M_S6.4 earthquake. *Chin. J. Geophys.* **2022**, *65*, 663–672.
7. Nathan, N.D.; MacKenzie, J.R. Rotational components of earthquake motion. *Can. J. Civ. Eng.* **2011**, *2*, 430–436. [CrossRef]
8. Yang, X.; Gao, F.; Chi, Q.; She, T.; Yang, L.; Wang, N. Study of strong earthquake rotational accelerometer based on a spoke-type mass-string system. *J. Nat. Disasters* **2015**, *24*, 37–45.
9. Gao, F.; Yang, X. Research on an active servo ultra-low frequency rotational accelerometer. *Earthq. Eng. Eng. Dyn.* **2018**, *38*, 171–178.
10. Qu, M.; Gao, F.; Yang, X.; Jia, X.; Yang, Q. Passive servo high-damping seismic rotational acceleration sensor. *Seismol. Geol.* **2018**, *40*, 231–239.
11. Takahashi, H.; Kan, T.; Nakai, A.; Takahata, T.; Usami, T.; Shimoyama, I. Highly sensitive and low-crosstalk angular acceleration sensor using mirror-symmetric liquid ring channels and MEMS piezoresistive cantilevers. *Sens. Actuators A Phys.* **2019**, *287*, 39–47. [CrossRef]
12. Nakashima, R.; Takahashi, H. Biaxial angular acceleration sensor with rotational-symmetric spiral channels and MEMS piezoresistive cantilevers. *Micromachines* **2021**, *23*, 507. [CrossRef]
13. Li, J.; Fang, J.; Du, M.; Dong, H. Analysis and fabrication of a novel MEMS pendulum angular accelerometer with electrostatic actuator feedback. *Microsyst. Technol.* **2013**, *19*, 9–16. [CrossRef]
14. Krishtop, V.G.; Agafonov, V.M. Technological principles of motion parameter transducers based on mass and charge transport in electrochemical microsystems. *Russ. J. Electrochem.* **2012**, *48*, 746–755. [CrossRef]
15. Chen, M.; Zhong, A.; Lu, Y.; Chen, J.; Chen, D.; Wang, J. A MEMS electrochemical angular accelerometer leveraging silicon-based three-electrode structure. *Micromachines* **2022**, *12*, 186. [CrossRef] [PubMed]
16. Liu, B.; Wang, J.; Chen, D.; Chen, J.; Xu, C.; Liang, T.; She, X. A MEMS based electrochemical angular accelerometer with integrated plane electrodes for seismic motion monitoring. *IEEE Sens. J.* **2020**, *20*, 10469–10475. [CrossRef]
17. Wang, S.H. Current status and applications of MEMS sensors. *Micro/Nano Electron. Technol.* **2011**, *48*, 516–522.
18. Kozlov, V.A.; Agafonov, V.M.; Bindle, J. Small, low-power, low-cost IMU for personal navigation and stabilization systems. In Proceedings of the 2006 National Technical Meeting of The Institute of Navigation, Monterey, AB, Canada, 1 January 2006.
19. Hurd, R.M.; Jordan, W.H. The principles of the solion. *Platin. Met. Rev.* **1960**, *4*, 42–47. [CrossRef]
20. Egorov, E.; Agafonov, V.; Avdyukhina, S.; Borisov, S. Angular molecular–electronic sensor with negative magnetohydrodynamic feedback. *Sensors* **2018**, *18*, 245. [CrossRef] [PubMed]

Article

Weak Capacitance Detection Circuit of Micro-Hemispherical Gyroscope Based on Common-Mode Feedback Fusion Modulation and Demodulation

Xiaoyang Zhang, Pinghua Li, Xuye Zhuang *, Yunlong Sheng *, Jinghao Liu, Zhongfeng Gao and Zhiyu Yu

School of Mechanical Engineering, Shandong University of Technology, Zibo 255000, China; zxyjiayyou@163.com (X.Z.)
* Correspondence: zxye@sdut.edu.cn (X.Z.); shengyunlong@sdut.edu.cn (Y.S.)

Abstract: As an effective capacitance signal produced by a micro-hemisphere gyro is usually below the pF level, and the capacitance reading process is susceptible to parasitic capacitance and environmental noise, it is highly difficult to acquire an effective capacitance signal. Reducing and suppressing noise in the gyro capacitance detection circuit is a key means to improve the performance of detecting the weak capacitance generated by MEMS gyros. In this paper, we propose a novel capacitance detection circuit, where three different means are utilized to achieve noise reduction. Firstly, the input common-mode feedback is applied to the circuit to solve the input common-mode voltage drift caused by both parasitic capacitance and gain capacitance. Secondly, a low-noise, high-gain amplifier is used to reduce the equivalent input noise. Thirdly, the modulator–demodulator and filter are introduced to the proposed circuit to effectively mitigate the side effects of noise; thus, the accuracy of capacitance detection can be further improved. The experimental results show that with the input voltage of 6 V, the newly designed circuit produces an output dynamic range of 102 dB and the output voltage noise of 5.69 nV/$\sqrt{\text{Hz}}$, achieving a sensitivity of 12.53 V/pF.

Keywords: hemispherical resonant gyro; common-mode feedback; micro-capacitor detection; low noise

Citation: Zhang, X.; Li, P.; Zhuang, X.; Sheng, Y.; Liu, J.; Gao, Z.; Yu, Z. Weak Capacitance Detection Circuit of Micro-Hemispherical Gyroscope Based on Common-Mode Feedback Fusion Modulation and Demodulation. *Micromachines* **2023**, *14*, 1161. https://doi.org/10.3390/mi14061161

Academic Editors: Aiqun Liu, Weidong Wang, Yong Ruan and Zai-Fa Zhou

Received: 12 April 2023
Revised: 23 May 2023
Accepted: 29 May 2023
Published: 31 May 2023

1. Introduction

With the characteristics of high reliability, high accuracy and long service life, hemispherical resonant gyroscopes (HRGs) have been widely used in multiple fields, such as aerospace, ship navigation and unmanned flight [1,2]. However, due to their high price, large size and high power consumption, the application of HRGs in civil scenarios is limited. In order to promote the application of hemispherical resonant gyro, researchers have designed Micro-compact hemispherical gyros, which inherits the excellent performance of traditional ones, but with a smaller size and lower cost. However, the hemispheric gyro signal is weak and its performance is seriously affected by external environmental factors such as temperature and humidity [3,4]. Therefore, it has become an important and urgent issue to suppress the noise in the capacitance detection process to enable the acquisition and feedback of weak signals generated by Micro-compact hemispherical gyros [5–7].

A hemispherical gyroscope can be equated to a dynamic pair of differentially varying capacitance, and by measuring the change in capacitance, information such as the angle/angular velocity can be measured. The output capacitance change of a gyroscope is typically $10^{-18} \sim 10^{-12}$ F, which requires the capacitance detection circuit to have high sensitivity, low noise and large dynamic range [8]. Designing a low-noise, high-accuracy capacitance detection circuit is an important way to improve the performance of a hemispherical gyroscope [9].

The main methods for detecting weak capacitive signals are voltage follower circuits, trans impedance amplifier circuits and charge amplifier circuits. Voltage follower circuits and trans impedance amplifier circuits are susceptible to parasitic capacitance, which reduces the accuracy of the detection circuit. Charge amplifiers provide a virtual ground, and are insensitive to parasitic capacitance and can achieve high gain and low output impedance. When extracting and detecting useful signals, they are easily drowned in noise such as parasitic capacitance or high-frequency interference signal, resulting in effective capacitance signals which cannot be read out. Wu et al. proposed a CMOS low-noise capacitive readout circuit, using a trans impedance amplifier as the core module of the circuit [10], adopting a correlated double sampling technique to suppress the noise and mismatch of the circuit. The noise and mismatch of the circuit are effectively suppressed. By using the full-differential circuit, the performance of the readout circuit in terms of resolution and dynamic range are improved. The circuit also achieves a capacitance resolution of 1.5 aF/Hz$^{1/2}$ and a dynamic range of over 100 dB, but it is susceptible to parasitic capacitance. Li et al. improved the dynamic performance index of the detection signal [11], which are collected from a hemispherical resonator gyro by introducing a correction network and system transfer function. However, a high-input buffer amplifier with impedance is selected for the preamplifier circuit, which is prone to thermal noise and sensitive to the drift of the common-mode voltage.

In order to improve the signal-to-noise ratio, reduce the effects of thermal noise and drift of common-mode voltage [12], in this paper, we employ a charge amplifier to reduce noise and improve the signal-to-noise ratio. In addition, common-mode feedback technology is applied to reduce the interference signal of the operational amplifier circuit, leading to the improvement in extracting and detecting useful signals. Thus, the accurate detection of the capacitor signal produced by hemispherical resonant gyro can be achieved. Moreover, modulation–demodulation and filtering technology are used to improve the accuracy and reduce the noise of the detection circuit.

This paper is conducted in the context of MEMS gyroscope. The article designed a Micro-hemispherical resonator gyro to carry out the experiment, so the method we propose is suitable for MEMS gyro. The research proposes a capacitance detection circuit, where three different means are utilized to achieve noise reduction, including input common-mode feedback, a low-noise high-gain amplifier, and proper modulator–demodulator and filter.

2. Device Architecture of Capacitive Displacement Detection

The operating mode of a hemispherical resonant gyroscope resonator is a four-antinode vibration mode (i.e., the number of waves in the annulus m = 2), and the finite element model of the hemispherical resonator is shown in Figure 1 [13]. The mode is synthesized from two simple merged modes, both 45° apart in the direction of the antinode axis, which are the driving and detecting modes of the hemispherical gyro, as shown in Figure 2a,b.

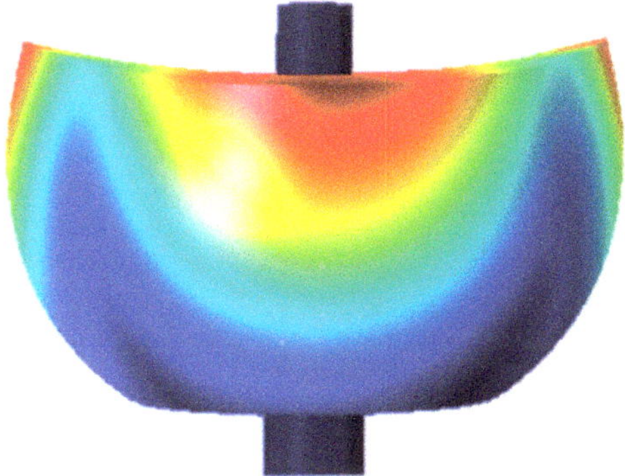

Figure 1. The finite element model of the hemispherical resonator.

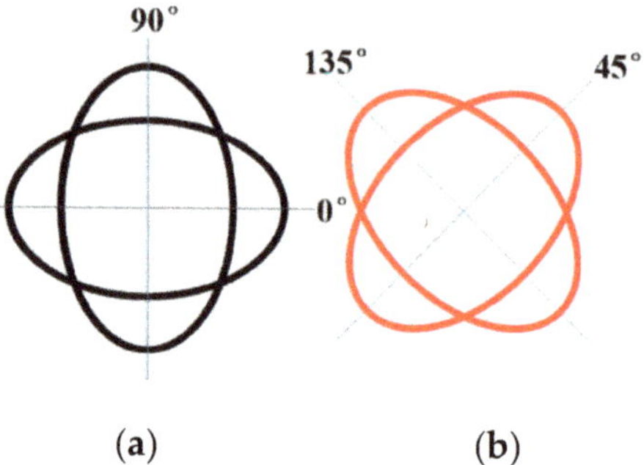

Figure 2. Driving mode and detection mode of miniature hemispherical gyroscope: (**a**) driving mode (**b**) detection mode.

The change in capacitance between the resonator shell and electrodes can be described by a schematic of detecting capacitance as shown in Figure 3. The differential capacitance consists of the hemispherical metal resonator, inner and outer detection electrodes distributed on the base. Moreover, variations in capacitance can be calculated accurately by calculating change in capacitance gap.

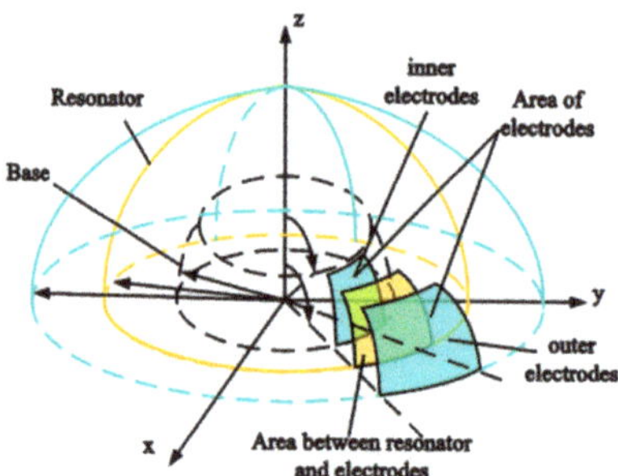

Figure 3. Schematic of detecting capacitance.

The principle of hemispherical resonator excitation and detection is shown in Figure 4, where the change in capacitance is converted into a voltage (V) by a test circuit, which in turn yields an angular velocity ω. When the hemispherical resonator of the gyroscope vibrates, the detection mode is excited by the Coriolis force, and the amplitude of the resonator detection mode is proportional to the magnitude of the input angular velocity, which is then detected by the detection electrode, and the signal is demodulated to obtain the input angular velocity. The amplitude of the detected mode is proportional to the input angular velocity.

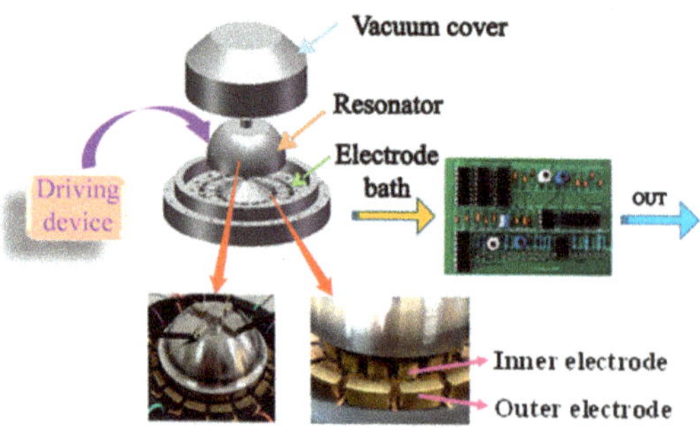

Figure 4. Schematic diagram of excitation and detection of hemispheric resonant gyroscope.

By applying an excitation signal to the resonator through the piezoelectric drive, the resonator will resonate, forming a four-antinode vibration, and the spacing between the resonator and electrode will change. So, a differential will be formed between the electrode and resonator, then a front-end amplifier circuit is connected from the chassis of the gyro structure to detect the capacitance change (ΔC). The differential electrode structure is shown in Figure 4, and the capacitance changes will be transformed by the test circuit into a voltage (V), which in turn gives the angular velocity (ω).

3. Circuit Design

3.1. Noise Analysis

Because of the small amplitude of the resonators' vibration, the gyro output signal is very weak. The readout circuit requirements are strict on noise, sensitivity and the compatibility of devices. The noises affecting the accuracy of the circuit are mainly the following: (1) thermal noise of the input MOSFET and feedback resistor; (2) $1/f$ noise of the input MOSFET; (3) reference noise generated by the reference power supply V_{ref}; (4) shot particle noise generated by the reverse bias diode leakage current. Figure 5 shows the block diagram of the Micro-miniature hemispherical resonant gyroscope detection circuit. The electrodes of different phases are excited by the piezoelectric drive, and a weak signal is generated. After, through the front-end amplifier circuit and C/V conversion circuit, the effective signal is obtained with a filter. Then, the effective signal is returned to the driving mode through the NI card and signal generator, thus forming a closed loop [14–16].

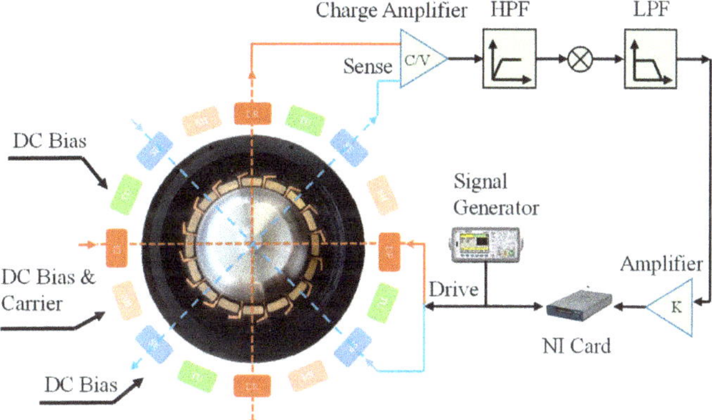

Figure 5. Block diagram of the hemispheric resonant gyroscope detection system.

The output voltage of the detection circuit charge amplifier V_F is

$$V_F = \Delta V \left(\frac{\partial C_1}{\partial t} - \frac{\partial C_2}{\partial t} \right) \times \frac{1}{C_{fp}s + 1/R_{fp}} \tag{1}$$

After the inverting terminal of the charge amplifier reaches the new negative reference voltage, V_{ref} is at the non-inverting end, the charge amplifier changes state again and the output is driven to the opposite supply rail voltage + V (sat). The capacitor gains a positive voltage across its plate and the charging cycle begins again. Thus, the capacitor is constantly charged and discharged, thus stabilizing the output of the charge amplifier signal. So, the charge amplifier circuit can reduce the effect of parasitic capacitance "C_p" at the output of the amplifier by using "V_{ref}" at the non-inverting terminal. For charge amplifiers, including differential amplifiers, there are non-inverting terminals connected to the reference voltage (V_{ref}) "+"and inverting terminals "$-$"connected to the output of the amplifier via a feedback capacitor (C_{fp}). The structure of the parasitic capacitance of the charge amplifier detection circuit is shown in Figure 6, where C_1 and C_2 are the capacitance of the gyroscope detect port, which are both C; C_p is the total parasitic capacitance of the circuit input, which mainly consists of the MOSFET gate capacitance C_{gd} and C_{gs}; R_{fp} and C_{fp} are the integral capacitance and the feedback resistor of the charge amplifier, respectively.

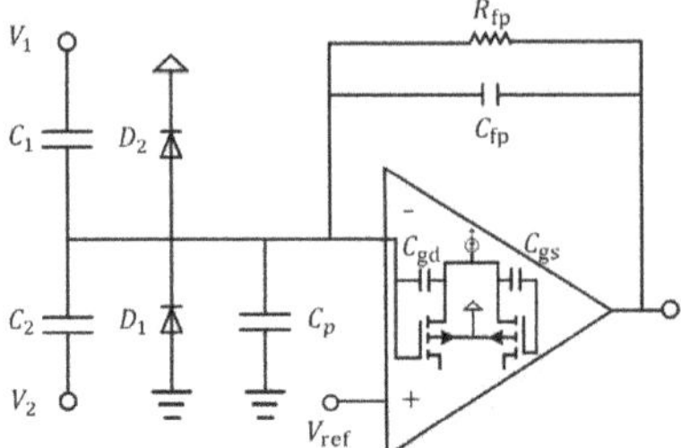

Figure 6. The parasitic capacitance structure of the detection circuit of the charge amplifier.

The reference voltage is V_{ref}, half of the voltage value of the power supply, and V_{O1} is the generated noise. The reference voltage noise at the output of the op-amp V_{O1} is

$$V_{O1} = \left(1 + \frac{C_p + C}{C_{fp}}\right) \times V_1 \tag{2}$$

The thermal noise, V_{O2}, is generated at the output of the operational amplifier, which can be defined as

$$V_{O2} = \frac{4KTR_{fp}}{2\pi f R_{fp}C_{fp} + 1} \tag{3}$$

The shot noise generated by the charge amplifier V_{amp} can be written as

$$V_{amp} = \frac{4KT}{\sqrt{2\mu C_{OX}(W/L)I_D}} + \frac{k}{C_{OX}WL} \times \frac{1}{f} \tag{4}$$

where W and L are the channel width and length parameters entered into the MOSFET, I_D is its bias current, K is the Boltzmann constant, T is the absolute temperature, μ is the carrier mobility, C_{OX} is the gate capacitance per unit area, k is the flicker noise factor and f is the frequency of circuit operating.

The shot noise at the output of the charge amplifier V_{O3} can be described as

$$V_{O3} = \left(1 + \frac{C_p + C}{C_{fp}}\right) \times \frac{4KT}{\sqrt{2\mu C_{OX}(W/L)I_D}} + \frac{k}{C_{OX}WL} \times \frac{1}{f} \tag{5}$$

The total output noise of the charge amplifier V_{out} is analyzed by the above calculation:

$$V_{out} = \left(1 + \frac{C_p + C}{C_{fp}}\right)V_1 + \frac{4KTR_{fp}}{2\pi f R_{fp}C_{fp} + 1} + \left(1 + \frac{C_p + C}{C_{fp}}\right)\frac{4KT}{\sqrt{2\mu C_{OX}(W/L)I_D}} + \frac{k}{C_{OX}WL}\frac{1}{f} \tag{6}$$

It shows clearly in Equation (6) that the output noise of charge amplifier is mainly affected by thermal noise and scintillation noise. At low frequencies, the circuit is affected by flicker noise. When the circuit operating frequency rises, the noise decays rapidly. The resonant frequency of the gyro is 6 kHz, and the noise level can be simplified as follows:

$$V_{out} = \left(1 + \frac{C_p + C}{C_{fp}}\right) \times V_1 + \left(1 + \frac{C_p + C}{C_{fp}}\right) \times \frac{4KT}{\sqrt{2\mu C_{OX}(W/L)I_D}} \tag{7}$$

As can be seen from Equation (7), $V_{out} \propto 1/C_{fb}$ when increasing the feedback capacitance of the charge amplifier, the noise at the output of the circuit will gradually decrease. However, as can be seen from Equation (1), $V_F \propto 1/C_{fb}$ increasing C_{fb} will lead to a reduction in the signal-to-noise ratio of the circuit. Therefore, to improve the accuracy of the circuit, the value of C_{fb} has to be reduced appropriately. In order to obtain a better overall performance of the circuit, the value of C_{fb} needs to be compromised according to the requirements of the gyroscope's using environment.

3.2. Operational Amplifier Design

The operational amplifier is the core module of the interface circuit, the noise, mismatch and limited DC gain of the operational amplifier will introduce errors into the capacitor–voltage conversion, resulting in poor detection accuracy [17,18]. Since the first-stage amplifier processes an extremely weak signal with a small output voltage amplitude, it is necessary to pay special attention to low noise and high gain when designing a gyro detection circuit. A folded cascode CMOS operational amplifier structure with PMOS tube input as shown in Figure 7 is used to achieve a large output voltage amplitude and wide dynamic range for the detection circuit; the output stage uses a class AB output structure to achieve a rail-to-rail output. The common-mode feedback compares the reference voltage with the common-mode value at the differential input of the main op-amp. The error value is sampled and delivered to the input feedback capacitor at the second phase, and when the next phase arrives, both pole plates of the input feedback capacitor are connected to the reference voltage. The common-mode feedback circuit introduced at the differential input reduces the parasitic capacitance and stabilizes the input common-mode voltage, while also effectively suppressing the thermal noise caused by the common-mode voltage drift and ensuring the stability of the input common-mode voltage of the main op-amp. Table 1 shows the technical index of the op-amp.

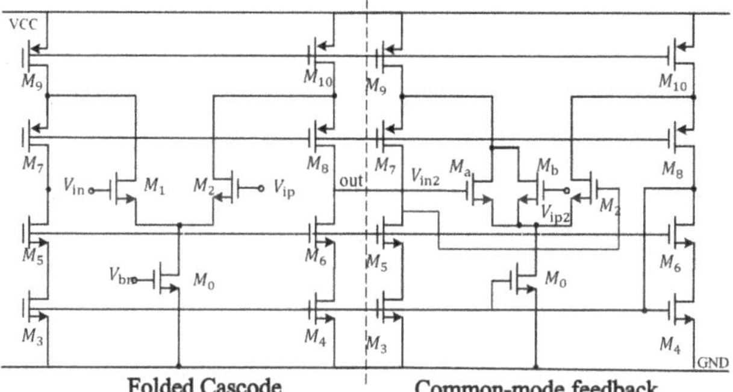

Figure 7. Structure diagram of operational amplifier.

Table 1. Technical index of operational amplifiers.

Performance Parameters	Indicators	Unit
Open-loop gain	102	dB
Gain bandwidth	18	MHz
Phase margin	45	°
Supply voltage	6	V
Equivalent input noise	6	nV/$\sqrt{\text{Hz}}$
Output amplitude	1~3	V

According to the technical specifications of the operational amplifier, and combined with Equations (1) and (7) to analyze and calculate the circuit, the integral capacitance of the charge amplifier is about 3.5 pF. Phase margin is a very important index in circuit design, which is mainly used to measure the stability of the negative feedback system and to predict the overshoot of the step response of the closed-loop system. For a good performance control system, the phase margin should be about 45 degrees. An amplifier with a small phase margin will respond longer, while an amplifier with a larger phase margin will take longer to rise to the final level of the voltage step, so the phase margin is about 45 degrees. The obtained open-loop amplitude and phase frequency curve of the amplifier are shown in Figure 8. Among them, the open-loop unit gain bandwidth of the operational amplifier is about 18 MHz and the open-loop gain is about 102 dB.

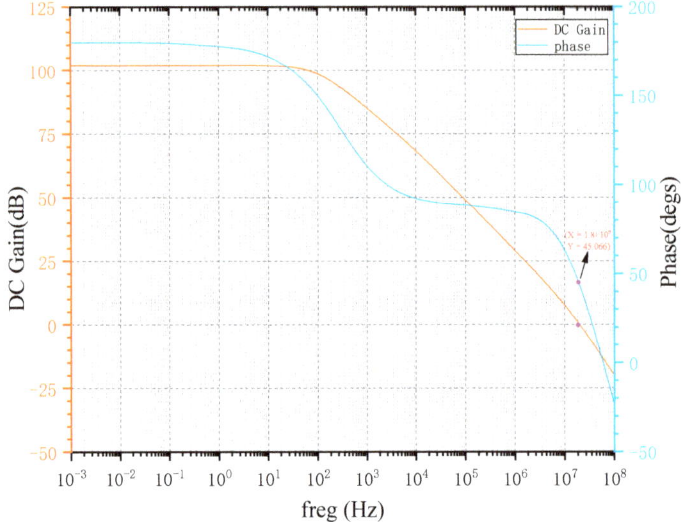

Figure 8. Open-loop amplitude frequency characteristic curve of operational amplifier.

4. Modulation–Demodulation Techniques

The modulation/demodulation technique is to modulate the current signal generated by the gyroscope via the stimulus signal. Specifically, it uses the dichotomous frequency relationship between the stimulus signal and the gyroscopic angular velocity signal to selectively amplify the signal. Using the filtering technique to process the signal of the first-stage amplification circuit, the band-pass filter outputs the frequency-doubled signal, retains the amplified angular velocity signal, and outputs it through the second-stage amplification circuit to complete the whole signal processing. Equation (7) shows that the circuit noise and frequency are inversely proportional to each other. When the frequency increases, the output noise of the circuit is reduced. Based on the characteristic of low noise in high frequency of the circuit, the modulation/demodulation technology could separate the signal from the noise and amplify an effective signal. Such operations effectively improve the accuracy of the readout circuit. The framework of the designed readout circuit for gyroscope capacitance is shown in Figure 9; the carrier signals are φ_1 and φ_2, and the demodulation switch is $\varnothing1, \varnothing2, \varnothing3$ and $\varnothing4$. The current signal generated by MEMS gyroscope is modulated by carrier signal, and the signal is selectively amplified by using the different working frequencies of carrier signal and gyroscope angular velocity signal, and the signal is processed by reasonable demodulation technology.

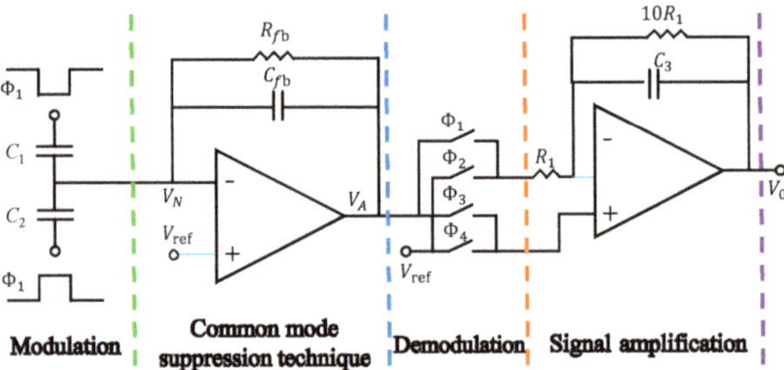

Figure 9. Design of gyroscope capacitance readout circuit.

Assume that the angular rate of gyroscopic detection is Ω and its magnitude is equal to

$$\Omega = \cos(\omega_\Omega t) \tag{8}$$

where ω_Ω is the rate of change in angular velocity.

The modulation of the small signal generated by the gyroscope with a carrier signal value of

$$f_n = \cos(\omega_n t) \tag{9}$$

where ω_d is the frequency of the carrier signal, then the output of the charge amplifier at the detection end is

$$V_A \infty 2 \times \Omega \times V \times f_n = 2\cos(\omega_\Omega t) \times \cos(\omega_d t) \times \cos(\omega_n t) \tag{10}$$

The signal is demodulated in two steps, starting with the carrier signal $\cos(\omega_n t)$ and then the drive signal $\cos(\omega_d t)$. The high-frequency signal is filtered using a filtering technique to obtain an output voltage proportional to the low-frequency angular velocity signal.

$$V_0 \infty \cos(\omega_\Omega t) \tag{11}$$

5. Analysis of Test Results

Using a sine signal of a certain amplitude generated by the signal generator as input, the waveforms of the positive and negative outputs of the circuit can be measured with an oscilloscope. The experimental test set-up is shown in Figure 10a, and the results of the read channel's transient response are shown in Figure 10b. In Figure 10b, the red curve represents the output voltage of the circuit before the improvement, from which it can be seen that the signal is messy and the noise is not suppressed. By using the methods of this research, the waveform changes from a messy 0.5 V to a 1 V sine wave. The green curve is acquired with the improved circuit, where the signal is sinusoidal, and the noise is effectively suppressed without obvious distortion.

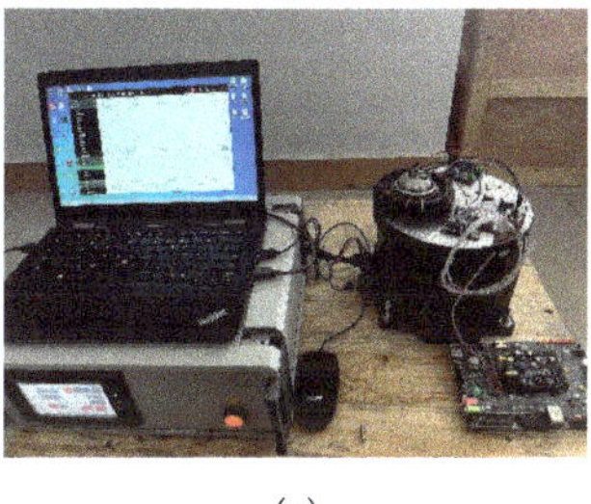
(a)

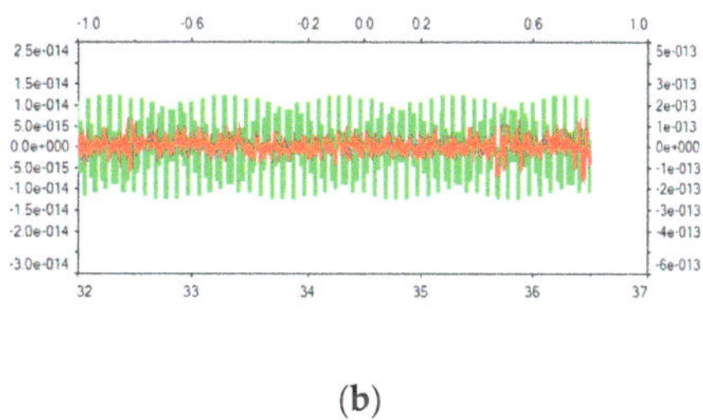
(b)

Figure 10. Transient response test diagram of readout circuit: (**a**) the experimental test set-up (**b**) the read channel's transient response.

The noise of the readout circuit is obtained using simulation for the voltage noise spectral density curve. The resonant frequency of MEMS gyroscope is 6 KHz, and the applied voltage is 4 V and 6 V, where the output voltage noise is 5.69 nV/$\sqrt{\text{Hz}}$ and the readout circuit noise ratio is approximately 0.27 in Figure 11. Then, the output noise of the amplifier designed in this article is carried out to obtain the noise power spectral density graph. The resonant frequency of gyro is close to 6 kHz, and the noise power spectral density is about -63.34 dBVrms/rtHz. The noise power spectral density of the amplifier output noise is shown in Figure 12.

The relationship among the circuit sensitivity, capacitance and voltage can be described in Figure 13. When the differential capacitance of gyroscope changes, the operational amplifier can output 1~3 V voltage. As the capacitance increases from -100 fF to $+100$ fF, the orange lines represent 10 experiments and 10 sets of data, and the red lines represent error lines, the sensitivity of the circuit maintains at 12.53 V/pF, the nonlinearity is 5.03×10^{-5}

and the operational amplifier circuit output voltage is 1.25 V. The conclusion is consistent with the technical index of the operational amplifier mentioned above.

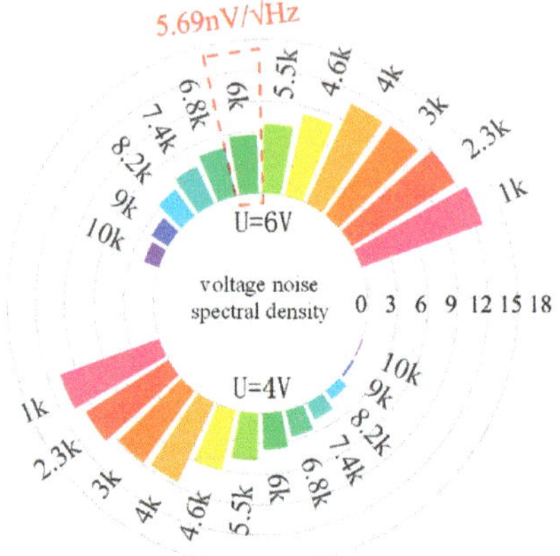

Figure 11. The spectrum density simulation diagram of voltage noise.

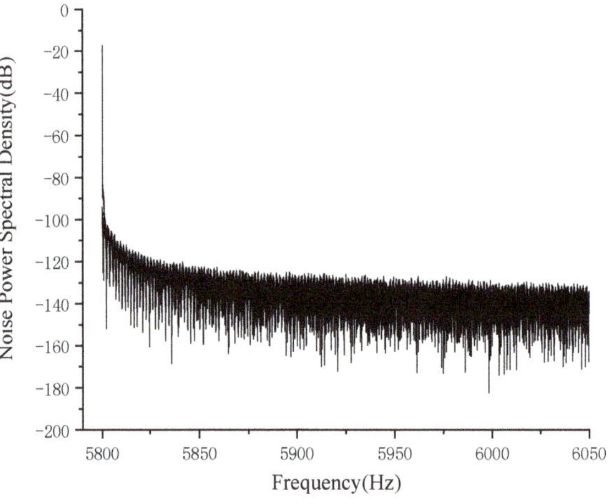

Figure 12. The noise power spectral density of the amplifier output noise.

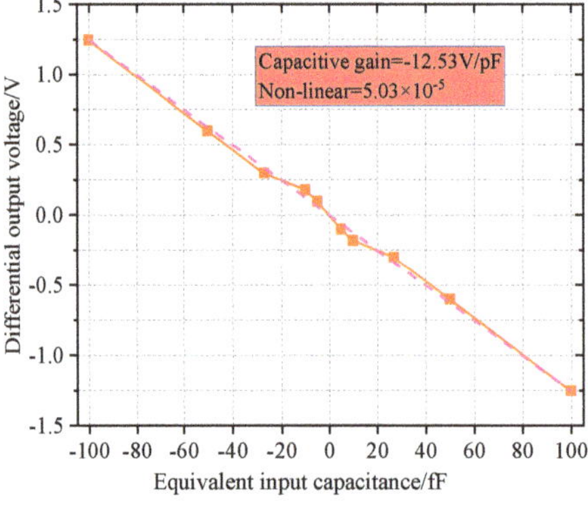

Figure 13. Sensitivity curve of input capacitance and output voltage.

In order to verify the accuracy of the circuit, we designed a small metal gyroscope; the frequency of the small metal gyro is 6 kHz with the biggest capacitance change of 0.5 pF. Figure 14 shows the measurement of frequency sweep for the four-antinode vibration mode tested at room temperature using network analyzer. Through the circuit-to-sweep frequency, the resonant frequency of the resonator is 6.12836 kHz with a quality factor of 2665. The results are consistent with the designed gyro frequency; it is proven that the circuit can be used as the weak capacitance detection circuit of MEMS gyroscope. Although the correctness of the circuit is verified, this method is complicated, and the process of testing requires a quiet space along with avoiding the introduction of any new noise.

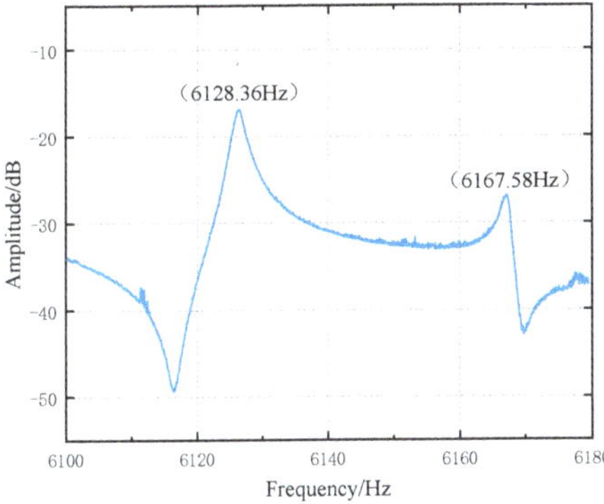

Figure 14. Frequency response of the resonant gyroscope.

In order to verify the long-term stability of the circuit, the designed circuit boards were tested at 20 degrees Celsius, 30 degrees Celsius and 40 degrees Celsius in Figure 15. At different temperatures, the test results are consistent with those in Figure 14, and the resonant frequency is 6.128 kHz. The test results show that the output signal of the circuit has no obvious change, which verifies the long-term stability of the design.

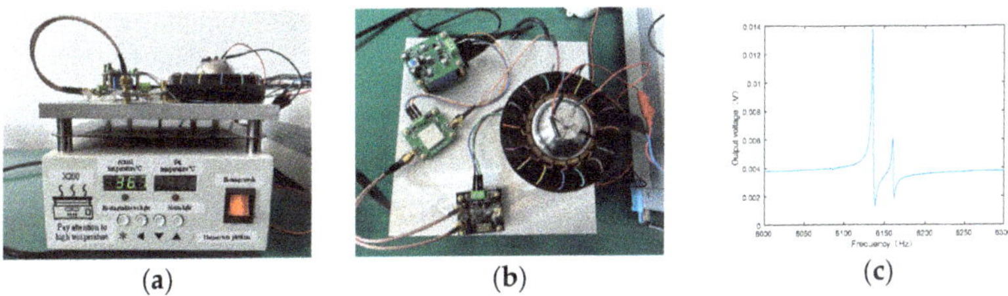

Figure 15. The long-term stability test of the readout circuit: (**a**) the experimental test set-up at different temperatures; (**b**) top view of the experimental test set-up; (**c**) frequency response of the resonant gyroscope.

The comparisons of the noise performance between different weak capacitance readout circuits are shown in Table 2. The circuit proposed in this article adopts common-mode feedback technology and modulation and demodulation technology, obtaining the output voltage noise of 5.69 nV/$\sqrt{\text{Hz}}$ and sensitivity of 12.53 V/pF. Wu et al. proposed a CMOS low-noise capacitive readout circuit, using a trans impedance amplifier as the core module of the circuit [10], adopting a correlated double sampling technique to suppress the noise and mismatch of the circuit. The noise and mismatch of the circuit are effectively suppressed. By using the full-differential circuit, the performance of the readout circuit in terms of resolution and dynamic range are improved. The circuit also achieves a capacitance resolution of 1.5 aF/Hz$^{1/2}$ and a dynamic range of over 100 dB, but it is susceptible to parasitic capacitance. Compared with reference [10] (1.5 aF/$\sqrt{\text{Hz}}$) and reference [19] (1.7 aF/$\sqrt{\text{Hz}}$), the capacitance resolution of the designed circuit is 0.1 aF/$\sqrt{\text{Hz}}$, which is improved by an order of magnitude. The sensitivity of the circuit introduced by reference [20] is 12.58 V/pF, which is slightly bigger than that of the circuit proposed in this article. In Ref. [17], an operational amplifier with a high power rejection ratio and low noise was designed, and the output level noise floor of the capacitive readout circuit was -117.14 dB, but the preamplifier circuit is chosen to be a transoresistor amplifier, which is easy to produce thermal noise and sensitive to the drift of the common-mode voltage. The output voltage noise ratio of our circuit is higher than the Northern General Electronics Group Limited circuit [16], which is increased by a factor of 0.27. The circuit designed here has the characteristics of low noise, small capacitive resolution and high sensitivity, which can be used in the field of weak capacitance detection.

Table 2. Performance comparison of different weak capacitor readout circuits.

	Output Voltage Noise (nV/$\sqrt{\text{Hz}}$)	Capacitive Resolution (aF/$\sqrt{\text{Hz}}$)	Sensitivity (V/pF)	Remarks
Institute of Electronics, Chinese Academy of Sciences	/	1.5	16.7	Reference [10]
East China Institute of Optoelectronic Integrated Devices	/	0.06	12.58	Reference [20]
Northern General Electronics Group Limited	7.82	0.116	/	Reference [16]
Amirkabir University of Technology	/	1.7	/	Reference [19]
This article	5.69	0.1	12.53	

6. Conclusions

With benefits including high precision, high reliability, long life and simple structure, the gyroscope has been widely used in many fields such as aerospace, autonomous vehicles and robotics. The rapid development of micromachines has put forward higher and higher requirements for the performance of subsequent readout circuits. However, the development of a high quality resonator does not make a high performance gyroscope. The interface and control systems circuit must be carefully designed to take full advantage of the resonator's capabilities and compensate for its non-idealities. The capacitance produced by the gyroscope changes very little. The weak capacitance changes are easily submerged in noise. Designing a weak capacitance detection circuit is of great significance to improve the performance of a gyroscope. The design and implementation of the electronic control and readout interface circuit is just as important as that of the resonator for defining the final performance of the gyroscope.

A weak capacitance detection circuit is designed to detect the capacitance generated by hemisphere gyro. In the circuit, the input common-mode voltage drift, which is caused by parasitic and gain capacitance, is effectively reduced via input common-mode feedback. A low-noise high-gain amplifier is applied to further suppress the equivalent input noise. Furthermore, the accuracy of the proposed circuit is significantly improved by modulator–demodulator and filter techniques, which work well in combating noise. Experimental results show that the output voltage noise, sensitivity, voltage and dynamic range of the proposed circuit are 5.69 nV/$\sqrt{\text{Hz}}$, 12.53 V/pF, 1.25 V and 102 dB, respectively. Such results demonstrate the effectiveness of the proposed circuit and verify that it can be used in the weak capacitance detection of Micro-hemisphere gyroscopes, and it provides a useful research value for the rapid development of micromachinery.

Author Contributions: Conceptualization, X.Z. (Xiaoyang Zhang) and P.L.; methodology, Z.G. and J.L.; software, Y.S. and X.Z. (Xuye Zhuang); validation, X.Z. (Xiaoyang Zhang) and Z.Y.; formal analysis, X.Z. (Xiaoyang Zhang) and Y.S.; data curation, and Z.G.; writing—original draft preparation, X.Z. (Xiaoyang Zhang) and X.Z. (Xuye Zhuang); writing—review and editing, X.Z. (Xiaoyang Zhang), X.Z. (Xuye Zhuang) and P.L.; visualization, X.Z. (Xuye Zhuang). All authors have read and agreed to the published version of the manuscript.

Funding: This work was supported by the Taishan Scholars Program of Shandong Province (No. tsqn201909108), the Key Research and Development Project of Zibo City under Grant No. 2020SNPT0088, the Natural Science Foundation of Shandong Province, China, Grant No. ZR2021MF042.

Conflicts of Interest: The authors declare no conflict of interest.

References

1. Chen, W.; Xiao, D.; Pu, J.; Li, K. Design and Implementation of Signal Detection Circuit for MEMS Vibration Sensor. *J. Sens. Microsyst.* **2021**, *40*, 63–66.
2. Zhang, J.; Li, P.; Zhuang, X.; Zhang, X.; Liu, J. Analysis of temperature drift error Compensation method for hemispherical resonant gyro. *Navig. Control* **2022**, *21*, 1–7+55.
3. Li, Z.M.; Lan, Z.; Jin, K.; Zhang, J.; Zhang, H. Parasitic capacitance adaptive suppression for fly-by-wire capacitive sensor readout circuits. *J. Xi'an Jiaotong Univ.* **2021**, *55*, 154–161.
4. Liu, S.; Li, M.; Meng, F.; Ma, A.; Su, M. Design of micro-capacitive sensor signal detection circuit. *Electron. Devices* **2020**, *43*, 613–619.
5. Qian, M. Research on Principle and Test Method of MEMS Gyroscope. *Integr. Circuit Appl.* **2017**, *34*, 72–77.
6. Zhuang, X.; Ding, J.; Cao, W.; Wu, M. Nondestructive testing of Silicon Hemisphere deep Cavity Structures. *Navig. Control* **2019**, *18*, 89–95.
7. Ataman, C.; Urey, H. Modeling and characterization of comb-actuated resonant microscanners. *J. Micromech. Microeng.* **2006**, *16*, 9. [CrossRef]
8. Gao, S.; Wu, J. Multi-electrode based hemispherical resonant gyro signal detection. *J. Harbin Eng. Univ.* **2008**, *29*, 475–477.
9. Yi, Y.; Song, A.; Li, H.; Leng, M. Design of micro-capacitive detection circuit for capacitive tactile sensors. *Instrum. Technol. Sens.* **2020**, *5*, 6–10.
10. Wu, Q.; Yang, H.; Zhang, C. A high-performance capacitive readout circuit for MEMS gyroscopes. *J. Instrum.* **2010**, *31*, 937–943.
11. Li, X.; Zheng, Y.B. Research on hemispherical resonant gyro signal detection and dynamic performance improvement method. *Navig. Control* **2015**, *14*, 32–33.
12. Zhang, M.; Wang, H.; Lv, D. Analysis and modeling of random errors in MEMS gyroscopes. *Electro-Opt. Control* **2022**, *29*, 68–71.
13. Zhang, W.; Chen, W.; Yin, L.; Di, X.; Chen, D.; Fu, Q.; Zhang, Y.; Liu, X. Study of the Influence of Phase Noise on the MEMS Disk Resonator Gyroscope Interface Circuit. *Sensors* **2020**, *20*, 5470. [CrossRef] [PubMed]
14. Tan, Q.; Li, L. Development history and key technologies of hemispherical resonant gyroscope. *Proc. Symp. Inert. Technol. Intell. Navig.* **2019**, 91–95.
15. Norouzpour-Shirazi, A.; Zaman, M.F.; Ayazi, F. A digital phase demodulation technique for resonant MEMS gyroscopes. *IEEE Sens. J.* **2014**, *14*, 3260–3266. [CrossRef]
16. Tang, X.; Long, S.; Liu, Y. High-precision low-noise MEMS gyroscope capacitive readout circuit. *J. Sens. Technol.* **2014**, *27*, 1191–1195.
17. Jiang, R.; Kong, D.; Li, Z.; Bao, L.; Lin, B.; Guo, P. Fully differential amplifier for MEMS micro capacitor detection. *Semicond. Technol.* **2009**, *34*, 181–184.
18. Tian, M. *Low-Noise Interface Circuit Design for Micro-Multi-Loop Resonant Gyroscopes*; Shanghai Jiaotong University: Shanghai, China, 2020.
19. Zargari, S.; Moezzi, M. A New Readout Circuit with Robust Quadrature Error Compensation for MEMS Vibratory Gyroscopes. *Circuits Syst. Signal Process.* **2022**, *41*, 3050–3065. [CrossRef]
20. Zhang, H.; Tong, Z.; Long, S.; He, K. A novel low-noise silicon micromechanical gyro-capacitor readout circuit. *Navig. Control* **2019**, *18*, 100–106.

Article

MEMS Differential Pressure Sensor with Dynamic Pressure Canceler for Precision Altitude Estimation

Shun Yasunaga [1], Hidetoshi Takahashi [2], Tomoyuki Takahata [3] and Isao Shimoyama [4,5,*]

1   Department of Electrical Engineering and Information Systems, Graduate School of Engineering, The University of Tokyo, 7-3-1 Hongo, Bunkyo-ku, Tokyo 113-8656, Japan; yasunaga@if.t.u-tokyo.ac.jp
2   Department of Mechanical Engineering, Faculty of Science and Technology, Keio University, 3-14-1 Hiyoshi, Kouhoku-ku, Yokohama 223-8522, Kanagawa, Japan; htakahashi@mech.keio.ac.jp
3   Department of Advanced Machinery Engineering, School of Engineering, Tokyo Denki University, 5 Senju Asahi-cho, Adachi-ku, Tokyo 120-8551, Japan; t-tkht@mail.dendai.ac.jp
4   Imizu Campus, Toyama Prefectural University, 5180 Kurokawa, Imizu 939-0398, Toyama, Japan
5   Hongo Campus, The University of Tokyo, 7-3-1 Hongo, Bunkyo-ku, Tokyo 113-8654, Japan
*   Correspondence: i-shimoyama@pu-toyama.ac.jp

Abstract: Atmospheric pressure measurements based on microelectromechanical systems (MEMSs) can extend accessibility to altitude information. A differential pressure sensor using a thin cantilever and an air chamber is a promising sensing element for sub-centimeter resolution. However, its vulnerability to wind and the lack of height estimation algorithms for real-time operation are issues that remain to be solved. We propose a sensor "cap" that cancels the wind effect and noise by utilizing the airflow around a sphere. A set of holes on the spherical cap transmits only the atmospheric pressure to the sensor. In addition, we have developed a height estimation method based on a discrete transfer function model. As a result, both dynamic pressure and noise are suppressed, and height is estimated under a 5 m/s wind, reconstructing the trajectory with an estimation error of 2.8 cm. The developed sensing system enhances height information in outdoor applications such as unmanned aerial vehicles and wave height measurements.

Keywords: MEMS; pressure sensor; wind effect; altitude estimation

Citation: Yasunaga, S.; Takahashi, H.; Takahata, T.; Shimoyama, I. MEMS Differential Pressure Sensor with Dynamic Pressure Canceler for Precision Altitude Estimation. *Micromachines* **2023**, *14*, 1941. https://doi.org/10.3390/mi14101941

Academic Editors: Weidong Wang, Yong Ruan and Zai-Fa Zhou

Received: 25 September 2023
Revised: 14 October 2023
Accepted: 16 October 2023
Published: 18 October 2023

1. Introduction

Altitude is one of the most important indices for localizing mobile agents and space mapping [1]. Conventional methods for measuring altitude include integrating accelerometer responses [2,3], calculations from radar measurements [4,5], and measuring the time of flight of light or sound reflecting off an object or the ground [6]. We can also utilize camera parallax [7,8] and global positioning system (GPS) information [9–11]. Estimating altitude from atmospheric pressure [2,3,12–15] is another method that can be used to acquire altitude information. Each method has its proper place and time of use, and among them, atmospheric pressure is a measurable physical quantity suitable for low-power altitude measurements in environments without reference for altitude, such as in the air or for ocean waves. Attitude estimation of unmanned aerial vehicles (UAVs) [6,13] and wave height measurements at sea [9,16], for example, can benefit from atmospheric-pressure-based measurements.

Previously, we developed a cantilever-type barometric sensor based on microelectromechanical systems (MEMSs) [17–19]. The sensor consists of a MEMS cantilever as the sensing element and an air chamber attached under the sensor chip. Altitude estimation based on this sensor can provide a resolution of less than 1 cm in height and has a fast response in the order of milliseconds [20], owing to the small mass of the sensing element. However, two factors remain to be solved before its practical application: vulnerability to wind and an estimation method for real-time operation.

Wind is a significant factor, particularly in outdoor applications, owing to the dynamic pressure created on the surface of the sensor. Although the source of altitude information is static pressure, it is impossible to distinguish it from dynamic pressure, which is inseparably superimposed in the measurement. According to Bernoulli's law, a wind speed of 1 m/s under standard conditions (0 °C, 1 atm) will produce a dynamic pressure of up to 0.6 Pa, which amounts to 5 cm of elevation near sea level and exceeds the altitude resolution of the sensor itself. Therefore, developing a system that can measure atmospheric pressure while suppressing the wind effect is crucial to accurately measuring air pressure [21,22].

Regarding the height estimation method, the working principle of the cantilever-type differential pressure sensor has been investigated to determine its transfer function [18,19]; however, an estimation method that allows real-time data processing involving height estimation has yet to be developed.

This study proposes a wind-tolerant sensor cap that can be attached to a barometric sensor for high-precision altitude measurements (Figure 1). Here, we especially focus on one-directional wind, which can be encountered, for example, on a cruising UAV that has limited load size and faces wind from the direction of travel. The sensor cap was designed and developed based on the airflow and pressure distribution on a spherical surface. By changing the positions of the holes on the sphere, we experimentally observed that dynamic pressure and noise corresponded to theoretical pressure distribution, which allowed the determination of an optimized sensor cap that suppressed the dynamic pressure and noise caused by wind. We also developed a height estimation method based on a digitalized transfer function model. Using this method, we demonstrated that height estimation is possible in the presence of wind.

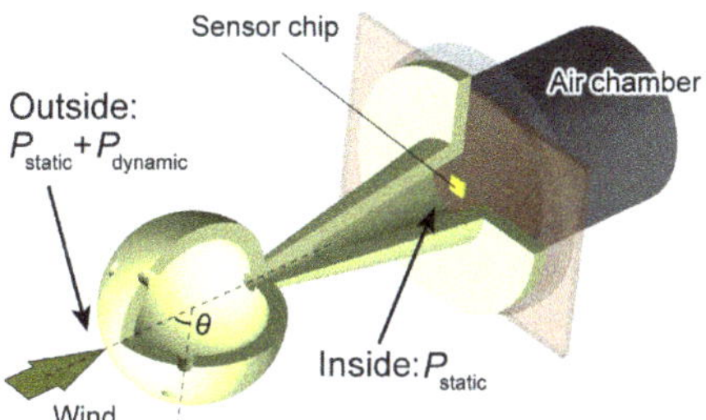

Figure 1. Concept of using a sensor cap to cancel dynamic pressure. At the bottom of the cap is a MEMS cantilever-type DPS attached to a printed circuit board and connected to an air chamber.

2. Materials and Methods

2.1. Theory and Requirements

We can estimate the altitude from a barometric pressure sensor signal because there is a one-to-one relationship between the height above sea level, h, and the atmospheric pressure, P. The height above sea level is calculated from the atmospheric pressure using [23]

$$P = P_0 \left(\frac{T_0}{T_0 + ch} \right)^{\frac{M_0 g}{Rc}}.$$

(1)

P_0 and T_0 are the atmospheric pressure and temperature at $h = 0$, respectively; c is the gradient of temperature per height (-6.5×10^{-3} K/m); M_0 is the average molar mass of air (28.8 g/mol); g is the acceleration due to the Earth's gravity (9.80 m/s^2); and R is the gas constant (8.31×10^3 m^2 g s^{-2} K^{-1} mol^{-1}). We can use the first-order Taylor expansion around $h = 0$,

$$\Delta P = -\frac{P_0 M_0 g}{RT_0} \Delta h,$$

(2)

for a displacement, Δh, near sea level, with less than a 1% error for 0–200 m. Although we cannot obtain absolute pressure from this simple relationship, it provides enough information in many situations, including the aforementioned possible applications.

However, the dynamic pressure caused by airflow around a pressure sensor influences the sensor's measurement. Dynamic pressure is a function of airflow velocity and is indistinguishably added to static (atmospheric) pressure. Moreover, the airflow around the sensor will cause local turbulence, which results in fast fluctuations (noise) in the sensor signal. Therefore, dynamic pressure and noise should be suppressed or compensated in outdoor applications or for movable platforms.

2.2. MEMS Pressure Sensor Element

General barometric sensors, called absolute pressure sensors (APSs), directly read atmospheric pressure values. Their operating principle is that the difference between the pressure of a reference air chamber and the external pressure will deform a diaphragm, and the extent of the deformation will indicate the atmospheric pressure [14]. Although an APS has a wide measurement range, it requires a robust structure that can withstand the maximum expected pressure, thus sacrificing sensitivity.

In this study, we used an MEMS differential pressure sensor (DPS) as a sensing element that was made sensitive without the need for robustness (Figure 2a). A thin cantilever, made using a silicon-on-insulator (SOI) wafer with a sub-μm-thick silicon device layer, sensed the differential pressure across a sensor chip. The differential pressure between the top and lower sides deformed the cantilever, which was detected as a change in the resistance of the piezoresistor, formed on the top side of the cantilever by doping. The cantilever legs were narrowed to concentrate the strain, resulting in a large resistance change.

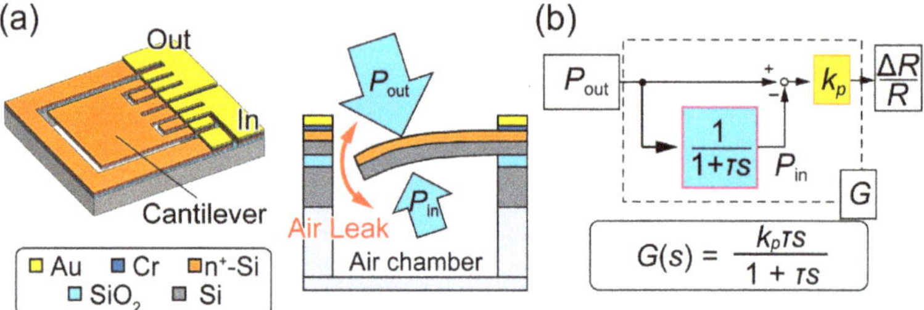

Figure 2. Concept of MEMS cantilever pressure sensor: (**a**) schematic and (**b**) transfer function model.

When an air chamber is attached to one side of the DPS, the cantilever will deform according to the differential pressure between the external (outside) pressure, P_{out}, and the internal chamber pressure, P_{in}. When P_{out} is atmospheric, the DPS will function as a pressure "change" sensor that can detect an ambient pressure change within a certain time. Hereafter, we define differential pressure as $\Delta P = P_{out} - P_{in}$. Unlike the diaphragm-type APS, the cantilever-type DPS allows airflow through the gap around the cantilever until the differential pressure vanishes (P_{in} reaches P_{out}). The combination of the DPS and the chamber has been identified as a first-order high-pass filter [19], with the transfer function from P_{out} to ΔP being $\tau s/(1 + \tau s)$, where τ is the time constant of the system (Figure 2b). The time constant is a measure of time it takes for P_{in} to reach P_{out}. The differential pressure is transduced to the fractional resistance change, $\Delta R/R$, at the sensor sensitivity rate, k_p, followed by conversion to a voltage and amplification with an amplifier gain, k_{amp}, before readout. Therefore, the transfer function of the sensing system, $G(s)$, is expressed as

$$G(s) = \frac{k_{amp} k_p \tau s}{1 + \tau s}. \tag{3}$$

The working principle of the DPS, in which the pressure inside the air chamber follows that of the outside by allowing air leakage through the gap, indicates that the time constant is a function of the chamber volume (a larger chamber requires a longer time constant and vice versa). This air leak limits the maximum differential pressure exerted on the cantilever surface. Unlike an APS, which would use a sealed chamber as a pressure reference and thus require a sensing structure sufficiently robust enough to withstand the unlimited differential pressure at the cost of sensitivity, the pressure reference of the DPS always stays close to the outside pressure, owing to the air leak at the gap, which allows for a thin, sensitive structure. We can also benefit from the thin structure in terms of its resonance frequency, which is high enough in the order of 10 kHz [20] to ensure accurate measurement of much slower pressure changes in height estimations.

2.3. Sensor Cap

To eliminate the dynamic pressure and noise caused by wind, we devised a sensor "cap". The dynamic pressure could be suppressed if we conducted a measurement at the exact point where the pressure is always equal to the static pressure regardless of the wind speed. A sphere is a good candidate for this purpose because of its omnidirectional shape and the predictable pressure distribution on its surface.

Considering the flow around a sphere, the pressure is the highest at the meeting point of the streamline, where the flow is stagnant, gradually decreasing to below atmospheric pressure as air flows faster around the surface. The theoretical pressure distribution, P, for laminar flow as a function of the angle from the streamline (θ) is given as [24]

$$P - P_0 = \frac{1}{2}\rho v^2 \left(1 - \frac{9}{4}\sin^2\theta\right), \tag{4}$$

where P_0 is the atmospheric (static) pressure, ρ is the air density, and v is the flow velocity. Although experiments have shown slightly different distributions than this theoretical calculation [25], in either case, the pressure around the sphere will reach the atmospheric pressure at a certain angle, θ_0, which is less than 90° (41.9° for laminar flow Equation (4)), regardless of the flow velocity. If we can extract and measure the pressure at this particular point, it will always be equal to the static pressure, even in the wind.

The proposed sensor cap has a spherical head with multiple evenly placed holes, each located at the same angle, θ, from the streamline, introducing pressure into the inside of the sensor cap. The effect of using multiple holes is that the cap averages the pressure from each hole to produce a less noisy pressure that is fed to the sensor. The spherical head is supported by a hollow pillar that separates the cap's head (where the pressure to be measured is taken in) from the rest of the sensor assembly, which would disturb the airflow if placed nearby. The pillar allows the pressure from the sphere to be transmitted to the MEMS cantilever pressure sensor.

2.4. Discrete Transfer Function Model for Height Estimation

We could obtain height information from the sensor output by applying the inverse transfer function $G^{-1}(s) = (1 + \tau s)/k_{\mathrm{amp}}k_\mathrm{p}\tau s$. We employed the bilinear z-transform to convert the transfer function of the continuous system into its discrete counterpart,

$$s = \frac{2}{T}\frac{1 - z^{-1}}{1 + z^{-1}}, \tag{5}$$

where T is the sampling period. The time constant for the discrete system is then

$$\frac{1}{\tau} = \frac{2}{T}\tan\left(\frac{T}{2\tau'}\right), \tag{6}$$

where τ' is the corrected time constant for the discrete system. Therefore, the transfer function for the discrete system is

$$G^{-1}(z) = \frac{(T + 2\tau') + (T - 2\tau')z^{-1}}{2k_{\mathrm{amp}}k_{\mathrm{p}}\tau'(1 - z^{-1})}. \tag{7}$$

The outside pressure change, P_{out}, at data point n (the sampling interval is T) can be calculated from the measured data, $S(n)$, using

$$P_{\mathrm{out}}(n) = P_{\mathrm{out}}(n-1) + \frac{1}{k_{\mathrm{amp}}k_{\mathrm{p}}}\left\{(S(n) - S(n-1)) + \frac{T}{\tau'}\frac{S(n) + S(n-1)}{2}\right\}. \tag{8}$$

This formula updates the pressure estimation.

3. Results

3.1. Fabrication of Sensing Elements

The fabrication process of the MEMS cantilever-type DPS followed that of previous research using an SOI wafer [20] (Figure 3a). First, a 300 nm-thick silicon device layer was doped using rapid thermal diffusion to form a piezoresistive layer. Chromium and gold layers were then deposited using evaporation and were patterned using photolithography and wet etching. The device layer was then etched to form a cantilever shape using deep reactive ion etching (DRIE). Finally, DRIE was used to etch the handle silicon layer from the back side, and the oxide layer was removed using hydrogen fluoride vapor. Figure 3b,c show the fabricated sensor chip. The cantilever was 100 μm in width and length and surrounded by a 3 μm-wide gap. Before the loading of the sensor chip, a through-hole (1 mm in diameter) was drilled through a printed circuit board (PCB). The fabricated sensor chip was then glued onto the PCB at the through-hole using epoxy resin and wire bonding.

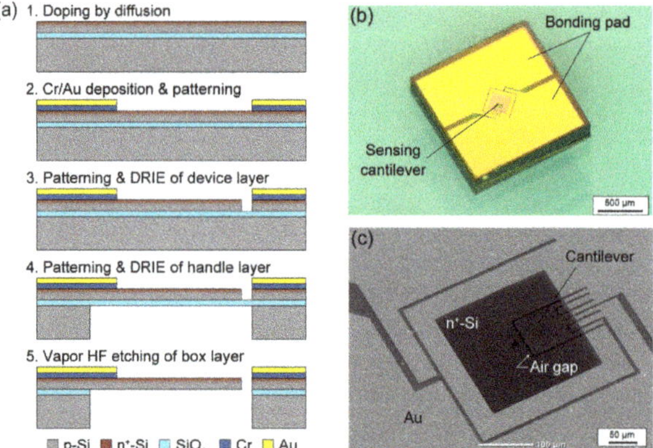

Figure 3. MEMS cantilever differential pressure sensor: (**a**) fabrication process, (**b**) overall appearance, and (**c**) SEM photograph.

3.2. Sensor System Development

Figure 4 illustrates the sensor cap we designed in this study. Sensing platforms for mounting pressure sensors, such as unmanned aerial vehicles and buoys, are limited in size and payload. In designing the sensor cap, the requirements are that (1) the size should be small enough not to interfere with the performance of the platform itself and (2) the region in which to measure the pressure should be small so that any gravity-induced pressure change is of the same order as the sensor's pressure resolution. Based on these requirements, we set the sphere size to be 10 mm in diameter, comparable to the height resolution of the DPS. A 20 mm-long hollow pillar that transmitted the pressure to the sensor chip separated the sphere from the rest of the sensor system, which disturbed the airflow. On the sphere, eight holes, each with a 0.5 mm diameter, were placed at a specific angle, θ, from the streamline to take in the pressure. The holes were sufficiently smaller than the surface area of the sphere so as not to disturb the flow but large enough for the pressure transmission. The number of holes was set to eight to obtain the effect of averaging signals for noise reduction while ensuring most of the flow would pass over the smooth surface of the sphere. We varied θ from $0°$ to $75°$ (note that the cap for when $\theta = 0°$ had only one hole) to experimentally obtain the optimal angle. These sensor caps were fabricated using a photopolymerizing 3D printer (EDEN260 V, Stratasys Ltd., Eden Prairie, Minnesota, USA; resin: Full-cure720), and the pillar of the sensor cap was painted black to eliminate the influence of light on the photosensitive piezoresistors.

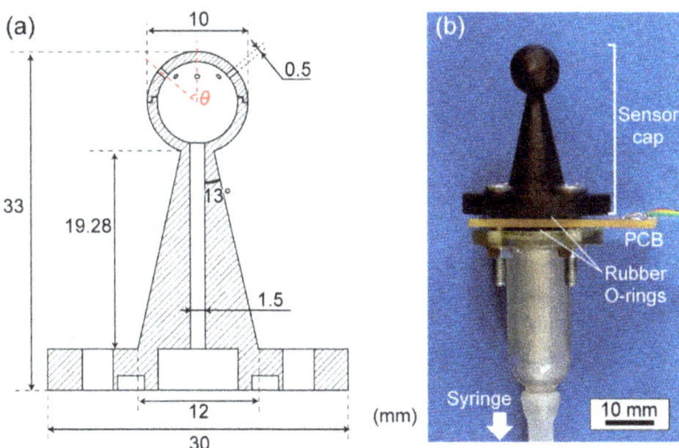

Figure 4. (**a**) Cap design and (**b**) fabricated sensor cap joined with PCB using O-rings and showing the tube and connector to a syringe to allow variable-volume tests.

If the holes on the sphere were too small or the volume inside the cap was too large, the transmission of pressure to the sensor chip would be delayed. The effect of the sensor cap on the delay of the pressure transmission was estimated as follows: The total inlet area of the eight circular holes was 150 times the area of the cantilever's surface and the gap combined (amounting to the maximum through-hole area on the sensor chip when the cantilever was fully bent). The sensor cap itself had an internal volume of 0.5 mL. Assuming that the flow was proportional to the cross-sectional area, the pressure inside the sensor cap reached equilibrium $150 \times (V \text{ [mL]}/0.5 \text{ [mL]})$ times faster than it would take for a V mL-chamber to reach the outside pressure by air leakage through the cantilever sensor chip. Therefore, the signal delay at the sensor cap was negligible, and we could assume that the pressure in the sensor cap was equal to the atmospheric pressure if the air chamber volume were comparable to or larger than 0.5 mL.

We prepared two types of air chambers: a rigid chamber with a fixed volume and a variable-volume chamber implemented using a syringe with a tube and a connector. The fixed-volume chamber and the connector with the variable volume were 3D-printed. The sensor cap, the PCB to hold the sensor chip in place, and the air chamber were connected by rubber O-rings and fastened using four screws. Figure 4b shows the completed sensor.

3.3. Sensor System Characterization

We first characterized the sensor by measuring the static and dynamic responses. According to Equation (3), the proposed sensor system is modeled based on two parameters: sensor sensitivity, k_p $(=(\Delta R/R)/\Delta P)$, and the time constant, τ [19]. We evaluated the sensor sensitivity by applying a constant differential pressure across the sensor chip and the time constant by using a step pressure input. The amplification gain, k_{amp}, in Equation (3) was 250 throughout this study, where the sensor output, as $\Delta R/R$, was measured using a one-gauge Wheatstone bridge circuit (bridge voltage = 1 V) and an instrumentation amplifier with a gain of 1000.

Figure 5a shows the measured $\Delta R/R$ versus the applied differential pressure, up to ± 40 Pa, using a pressure calibrator (KAL100; Halstrup-Walcher GmbH, Kirchzarten, Germany). The gradient in the plot corresponds to the sensor sensitivity, k_p. The result shows a linear relationship in the measured range, where $k_\mathrm{p} = -1.80 \times 10^{-4}$ Pa^{-1}.

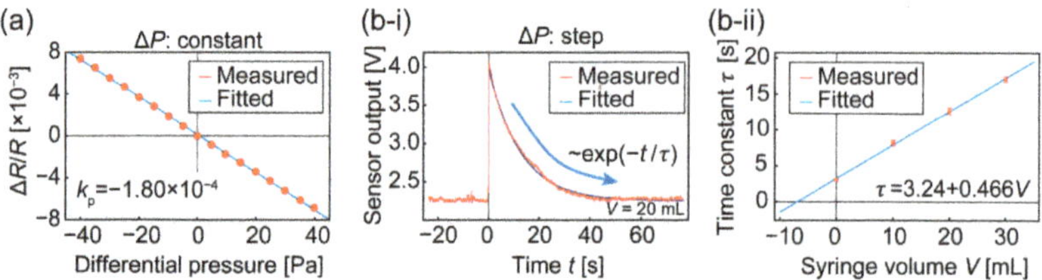

Figure 5. System calibration curves. (**a**) Sensitivity of sensor chip and (**b-i,b-ii**) dynamic characteristics of the system (time-series response (**b-i**) and obtained time constant vs. chamber volume (**b-ii**)). Error bars in (**b-ii**) correspond to the mean value $\pm$ the standard deviation for ten measurements.

The time constant, τ, of the system was measured using a step input and several chamber volumes (using the syringe arrangement). The step input was emulated in the measurement by manually opening a valve that had confined the sensor system under compressed air. Theoretically, the sensor system modeled in Equation (3) responds to a step input of $P_\mathrm{out}(t) = P_\mathrm{step}u(t)$ exponentially as $k_\mathrm{p}P_\mathrm{step}\exp(-t/\tau)$, where

$$u(t) = \begin{cases} 0 \ (t < 0) \\ 1 \ (t \geq 0) \end{cases}. \tag{9}$$

The measured responses shown in Figure 5b-i agree with this theory. We obtained the time constant by fitting an exponential curve to the response 0.2 s after opening the valve to remove the effect of unideal step input. The results for several chamber volumes, shown in Figure 5b-ii, were in a proportional relationship that was compatible with the working principle that air leakage allows inside pressure to eventually become equal to outside pressure [19]. Note that the horizontal axis represents the volume of the syringe, excluding that of the tube and connector. The intercept of the horizontal axis should correspond to the summed volume of the tube and the connector. These results indicate that the sensor system functioned as a differential pressure sensor, obeying transfer function model Equation (3).

3.4. Evaluation of the Cap

Using the characterized DPS and the experimental setup illustrated in Figure 6a, we evaluated the performance of the sensor cap. The sensor cap was mounted on the PCB sensor assembly, and the wind was applied using a direct-current (DC) fan placed upstream of the sensor cap. The chamber was the tube-connected syringe ($V = 100$ mL), whose time constant was $\tau = 50$ s. Figure 6b shows the measured responses when a wind velocity of $v = 8$ m/s was applied, kept on, and stopped for various θ values. When $\theta = 0°$, with the pressure-taking-in hole only at the stagnant point, the sensor output dropped as the DC fan was turned on ($t = 0$ s), corresponding to an increase in P_out (because of the negative k_p). The signal then increased, with noise toward the origin, as P_in followed P_out while the wind was present. After 30 s, when the DC fan was turned off and P_out became the atmospheric pressure, P_in was higher than P_out, resulting in a positive sensor output higher than at the initial state.

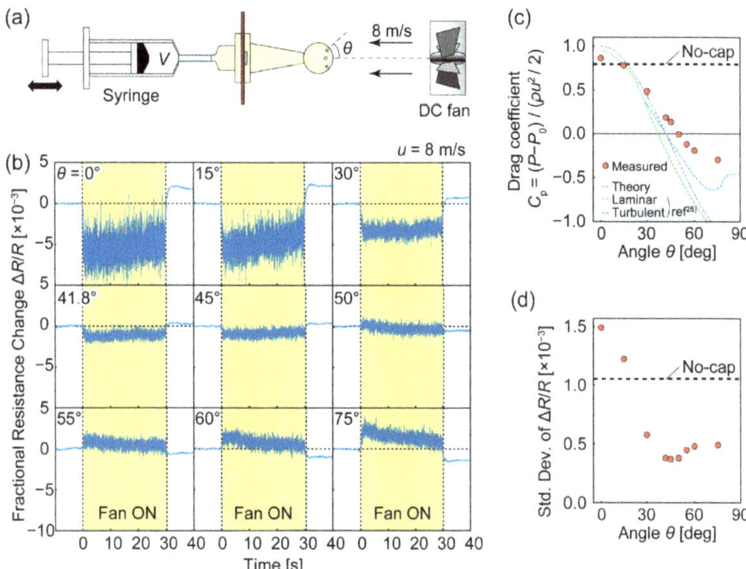

Figure 6. Evaluation of sensor cap: (**a**) experimental setup, (**b**) sensor output as the fan was turned on and off, (**c**) effect of dynamic pressure, and (**d**) noise vs. angle of incidence. In (**c**), the gray dashed line corresponds to the theoretical distribution on a sphere (Equation (4)), and the blue and green dashed lines correspond to experimental values obtained for laminar and turbulent flows, respectively [25].

The overall response to a change in wind speed decreased as θ increased to 50° and intensified again in the opposite direction for an even larger θ. To evaluate the effect of dynamic pressure, we fitted a line to the response $t = 2$–12 s and calculated its extrapolation to $t = 0$ s as the initial differential pressure, ΔP. Figure 6c shows the drag coefficient obtained from ΔP divided by the kinetic energy, $\rho v^2 / 2$, where ρ is the air density. The theoretical pressure distribution (Equation (4)), as well as the reported experimental values [25], are also plotted. As a result, the measured dynamic pressure as a function of θ followed a sinusoidal curve, corresponding to the theoretical prediction. The sensor cap with $\theta = 50°$ reduced the dynamic pressure the most successfully; for it, the wind effect was 0.90% of the result of the no-cap condition.

We also evaluated the noise in the airflow using this measurement. The dynamic pressure effect was compensated by subtracting the fitting line obtained from the $t = 2$–12 s response; then, the standard deviation was calculated as the noise. The result in Figure 6d is distributed similarly to the dynamic pressure amplitude in Figure 6c, implying the relationship between noise and dynamic pressure. The noise was lowest at $\theta = 45°$, with 35% of that measured in the no-cap condition, and the $\theta = 50°$ cap, which reduced the dynamic pressure the most, suppressed the noise to 36%. From the viewpoints of both dynamic pressure reduction and noise suppression, we concluded that $\theta = 50°$ was the best placement of the hole to take in pressure.

3.5. Height Estimation Method and Demonstration

We evaluated the height estimation performance of the sensor system using the $\theta = 50°$ cap and a 5 m/s wind. The measurement setup with the DC fan was loaded on a platform, which was winched up and down approximately 1 m at a rate of 0.17 m/s (Figure 7a). The same sensor chip and readout circuits used in the measurements thus far were employed, and the chamber volume was set to 5 mL. The atmospheric conditions were $P_0 = 101$ kPa and $T_0 = 14.8\,^\circ$C, which reduced Equation (2) to

$$\Delta P\,[\text{Pa}] = -12\,[\text{Pa/m}] \times (\Delta h\,[\text{m}]). \tag{10}$$

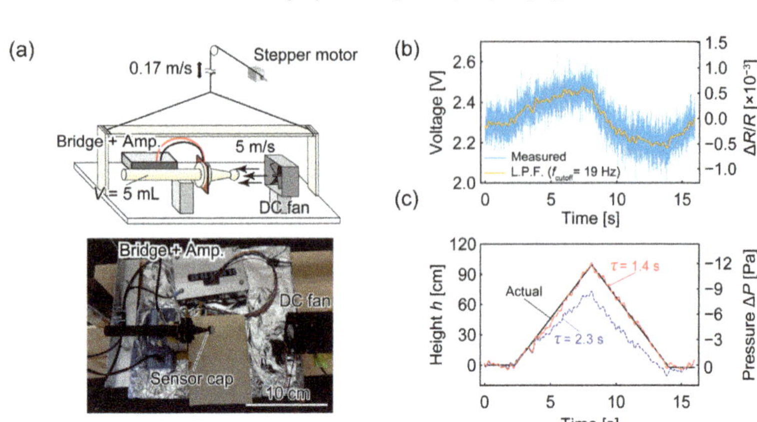

Figure 7. Height estimation under 5 m/s wind conditions: (**a**) experimental setup, (**b**) measured and low-pass filtered (L.P.F.) sensor output, and (**c**) height estimations vs actual height profile.

Figure 7b shows the sensor output measured at 500 Hz and the processed signal using a low-pass filter with a cutoff frequency of 19 Hz (see Discussion for the required sampling frequencies based on applications). The low-pass-filtered data were then processed using Equation (8) to obtain the pressure change, as shown in Figure 7c. The solid black line indicates the actual displacement of the platform. The blue dashed line is the estimated height using the time constant $\tau = 2.3$ s, calculated from Figure 5b-ii and the chamber volume of 5 mL, and the red dashed line shows the results using $\tau = 1.4$ s. The estimated profiles for the two time constants were triangular, similarly to the actual profile obtained as the platform moved. However, the estimated displacement with $\tau = 2.3$ s was smaller overall. When we recalculated the trajectory with different time constant values, $\tau = 1.4$ s reproduced the actual trajectory most accurately; for this, the root mean square of the estimation error was 2.8 cm.

It is implausible to attribute the difference in the time constant to sensor fabrication error because $\tau = 1.4$ s corresponds to a chamber volume of 3.0 mL, which is only 61% of the actual volume used, according to the proportional relationship between the chamber volume and the time constant obtained in Figure 5b. We can reasonably assume that the time constant in this experiment was shorter than the value obtained from Figure 5b, owing to air leakage between the PCB and the air chamber. In this experiment, the air chamber was a fully 3D-printed rigid chamber instead of the syringe attached via a 3D-printed connector. Although both the rigid chamber and the connector were 3D-printed, the surface of the connector was smoother because of wearing after many attach/detach operations and the orientation during printing. In general, photopolymerizing 3D printers produce rougher surfaces on the vertical face than on the horizontal, and the connector was printed horizontally whereas the rigid chamber was printed vertically. The rougher surface of the rigid chamber, which was not completely sealed by the rubber O-ring, leaked air through an alternate route than the supposedly only path for air to move in and out of the chamber, accelerating the equalization of the pressure inside and outside of the chamber. However, when air leaks through the rougher surface of the rigid chamber caused air leakage between the air chamber and the PCB in addition to the usual path across the sensor chip, the flow rate to equalize the differential pressure increased, therefore shortening the time constant.

To compare the noise unrelated to elevation for different wind speeds, the standard deviation of the measurement under 8 m/s without vertical movement was 2.84 cm after a 19 Hz low-pass filter, while for 5 m/s, it was 2.71 cm. This indicates that the effect of the wind-related fast-fluctuating noise was not dominant in the estimation, proving the effectiveness of the sensor cap in reducing the noise.

4. Discussion

In the sensor cap evaluation, we obtained $\theta = 50°$ as the best angle from streamline to place the hole to take in the pressure. This value is larger than the predicted theoretical and previously reported values [25], which might be attributable to the existence of the sensor structure after the flow moves around the sphere. Unlike an isolated sphere, the sensor assembly includes the cap's support (for screw fastening), the PCB, and the air chamber, which cause dynamic pressure. Although the pillar separating the spherical cap head from the rest of the assembly will ease their effect, the pressure distribution around the sphere will likely be higher than that of an isolated sphere, resulting in a larger θ_0.

A height estimation accuracy of 2.8 cm is comparable to the height resolutions of radiofrequency-based ultrawideband sensors [26,27], ultrasonic sensors [6], and GPS velocity estimation [11]. We have summarized the comparison to other sensing techniques in Table 1. Regarding the signal delay discussed in Section 3.2, the time constant for a 5 mL chamber, obtained in Section 3.5, was on the level of 1 s. Therefore, the delay due to the proposed sensor cap should have been on the level of 1 ms or less. This value is sufficiently low, considering the frequency characteristics of the cantilever differential-pressure sensor, which ensures accurate measurements at up to 200 Hz [20]. Therefore, the most significant factor limiting the sampling bandwidth is the low-pass filter used to eliminate measurement noise. The bandwidth should be chosen according to the specific requirements of the application and the expected noise level. For example, some signals for UAVs are sampled at 10–20 Hz [28,29], and the frequency band of maritime waves is 0.03–0.5 Hz [10,11].

Table 1. Performance compared among height-measuring techniques.

Method	Range	Differential Accuracy
Ultrasound ToF [30]	30 m	2 cm
RF ToF [27]	40 m	2.1 cm
Radar [31,32]	≈200 m	<5 cm
Absolute Pressure Sensor [33]	−730 m–9500 m [1]	23 cm [2]
This Work	200 m [3]	2.8 cm

[1] Values of 30–110 kPa, converted with Equation (1). [2] A value of 3.9 Pa, converted with Equation (10). [3] A 1% error range when using Equation (2); structurally unlimited.

As indicated by a previous study involving sensor fusion [18], MEMS cantilever pressure sensors should be used together with absolute pressure sensors to achieve the long-term stability of the sensing system. Although cantilever pressure sensors are capable of high sensitivity, they suffer from estimation drift because they are essentially a high-pass filter, and integral calculations are necessary in the estimation process. In contrast, absolute pressure sensors, including the diaphragm type, have low sensitivity but remain stable over time. By combining the proposed sensor system with the low-pass-filtered signal of an absolute pressure sensor, both long-term stability and high sensitivity can be achieved. This also provides the information of absolute altitude that might be required in some applications. In this research, we focused on the performance of the system using MEMS cantilever pressure sensors and performed a demonstration over a short period in which long-term stability was not required; however, the sensor cap can be applied to absolute pressure sensors as well. Thus, long-term precision measurements of pressure in different environments should be possible using the sensor cap.

The time constant of the sensor should be chosen according to the expected rate of pressure change and the measurable range of the cantilever in use. A long time constant, which can be realized by a large chamber or by a narrow gap around the cantilever, will have less integration-derived drift in its estimation because the pressure in the chamber will be maintained for a long period of time. However, it can cope with a smaller range of pressure change because the differential pressure must be within the linearly measurable range of the DPS. Likewise, a short time constant can be used even in a fast movement but can have more impact on the drift in estimation. A rough criterion for the choice of time constant is the linearly measurable range divided by the expected rate-of-pressure change (obtained from the maximum vertical speed and Equation (2)).

In outdoor applications, it is unrealistic to assume that wind blows in a fixed direction. To cope with this omnidirectionality, several of the proposed sensor systems can be placed in multiple directions so that the appropriate sensor can be chosen for estimation. Because the proposed sensor cap is on the level of centimeters, loading multiple sensor systems on small platforms should be possible. However, it is impossible to cover all spherical angles using a limited number of sensor systems. In this regard, we can expect that because the holes on the sphere have rotational symmetry, with evenly placed holes, the measurement will become robust against tilted streamlines, although we have not yet verified this. If a hole faces the airflow at a smaller angle, θ_0, and experiences a positive dynamic pressure, the effective angle of the hole on the opposite side will be greater than θ_0 and, therefore, will experience a countering negative dynamic pressure because the pressure distribution near $\theta = \theta_0$ will nearly be linear. After the pressures are combined within the sensor cap, the dynamic pressure should still be canceled. However, when the angle grows too much, the pressure distribution will no longer be linear. In the case of imperfectly turbulent flow, for example, the flow separates from the surface at approximately $\theta = 80°$ [25], giving a sudden change in pressure. This situation corresponds to a tilt angle of 30° from the streamline for the 50° cap, which might be the maximum acceptable angle.

Another factor to be considered in the outdoor application is measurement robustness against changing conditions. The cantilever DPS made of crystalline silicon and gold has robust durability but can be affected by temperature because of the temperature-dependent resistivity of doped silicon [20,34]. To avoid this dependency, a temperature-compensating resistor can be introduced on the same sensor chip and incorporated in the Wheatstone bridge [35].

5. Conclusions

To eliminate the wind effect in pressure measurements using a cantilever-type MEMS sensor, we developed a 10 mm spherical head cap with eight holes and a supporting hollow pillar. When each of the holes on the sphere faced the streamline at 50°, the dynamic pressure and noise were suppressed to 0.9% and 36% of those measured for the no-cap condition, respectively, considering an 8 m/s wind. We also developed a height estimation formula and demonstrated its accuracy based on a reconstructed height profile of 1 m of vertical movement under 5 m/s wind conditions, yielding an estimation error of 2.8 cm. We expect that the proposed sensing system will contribute to expanding the availability of pressure-based height information in outdoor applications that require high sensitivity.

Author Contributions: Conceptualization, S.Y. and H.T.; methodology, S.Y., H.T., T.T. and I.S.; software, S.Y.; validation, S.Y. and H.T.; formal analysis, S.Y.; investigation, S.Y.; resources, H.T.; data curation, S.Y.; writing—original draft preparation, S.Y.; writing—review and editing, H.T., T.T. and I.S.; visualization, S.Y.; supervision, I.S.; project administration, H.T. and T.T.; funding acquisition, H.T., T.T. and I.S. All authors have read and agreed to the published version of the manuscript.

Funding: This research received no external funding.

References

1. Bandari, S.; Devi, L.N. An Optimal UAV Height Localization for Maximum Target Coverage Using Improved Deer Hunting Optimization Algorithm. *Int. J. Intell. Robot Appl.* **2022**, *6*, 773–790. [CrossRef]

2. Tanigawa, M.; Luinge, H.; Schipper, L.; Slycke, P. Drift-Free Dynamic Height Sensor Using MEMS IMU Aided by MEMS Pressure Sensor. In Proceedings of the 2008 5th Workshop on Positioning, Navigation and Communication, Hannover, Germany, 27 March 2008; pp. 191–196.
3. Son, Y.; Oh, S. A Barometer-IMU Fusion Method for Vertical Velocity and Height Estimation. In Proceedings of the 2015 IEEE Sensors, Busan, Republic of Korea, 1–4 November 2015; pp. 1–4.
4. Dogru, S.; Marques, L. Pursuing Drones with Drones Using Millimeter Wave Radar. *IEEE Robot Autom. Lett.* **2020**, *5*, 4156–4163. [CrossRef]
5. Rai, P.K.; Kumar, A.; Khan, M.Z.A.; Soumya, J.; Cenkeramaddi, L.R. Angle and Height Estimation Technique for Aerial Vehicles Using MmWave FMCW Radar. In Proceedings of the 2021 International Conference on COMmunication Systems and NETworkS, COMSNETS 2021, Bangalore, India, 5–9 January 2021; pp. 104–108.
6. Wilson, A.N.; Kumar, A.; Jha, A.; Cenkeramaddi, L.R. Embedded Sensors, Communication Technologies, Computing Platforms and Machine Learning for UAVs: A Review. *IEEE Sens. J.* **2022**, *22*, 1807–1826. [CrossRef]
7. Jung, I.-K.; Lacroix, S. High Resolution Terrain Mapping Using Low Attitude Aerial Stereo Imagery. In Proceedings of the Ninth IEEE International Conference on Computer Vision, Nice, France, 13–16 October 2003; Volume 2, pp. 946–951.
8. Meingast, M.; Geyer, C.; Sastry, S. Vision Based Terrain Recovery for Landing Unmanned Aerial Vehicles. In Proceedings of the Proceedings of the IEEE Conference on Decision and Control, Nassau, Bahamas, 14–17 December 2004; Volume 2, pp. 1670–1675.
9. Qin, L.; Li, Y. Significant Wave Height Estimation Using Multi-Satellite Observations from GNSS-R. *Remote Sens.* **2021**, *13*, 4806. [CrossRef]
10. Herbers, T.H.C.; Jessen, P.F.; Janssen, T.T.; Colbert, D.B.; MacMahan, J.H. Observing Ocean Surface Waves with GPS-Tracked Buoys. *J. Atmos. Ocean Technol.* **2012**, *29*, 944–959. [CrossRef]
11. Doong, D.J.; Lee, B.C.; Kao, C.C. Wave Measurements Using GPS Velocity Signals. *Sensors* **2011**, *11*, 1043–1058. [CrossRef] [PubMed]
12. del Rosario, M.B.; Redmond, S.J.; Lovell, N.H. Tracking the Evolution of Smartphone Sensing for Monitoring Human Movement. *Sensors* **2015**, *15*, 18901–18933. [CrossRef]
13. Sabatini, A.M.; Genovese, V. A Sensor Fusion Method for Tracking Vertical Velocity and Height Based on Inertial and Barometric Altimeter Measurements. *Sensors* **2014**, *14*, 13324–13347. [CrossRef]
14. Kumar, S.S.; Tanwar, A. Development of a MEMS-Based Barometric Pressure Sensor for Micro Air Vehicle (MAV) Altitude Measurement. *Microsyst. Technol.* **2020**, *26*, 901–912. [CrossRef]
15. Li, Y.; Gao, Z.; He, Z.; Zhang, P.; Chen, R.; El-Sheimy, N. Multi-Sensor Multi-Floor 3D Localization With Robust Floor Detection. *IEEE Access* **2018**, *6*, 76689–76699. [CrossRef]
16. Grevemeyer, I.; Herber, R.; Essen, H. H. Microseismological Evidence for a Changing Wave Climate in the Northeast Atlantic Ocean. *Nature* **2000**, *408*, 349–352. [CrossRef] [PubMed]
17. Minh-Dung, N.; Takahashi, H.; Uchiyama, T.; Matsumoto, K.; Shimoyama, I. A Barometric Pressure Sensor Based on the Air-Gap Scale Effect in a Cantilever. *Appl. Phys. Lett.* **2013**, *103*, 4824027. [CrossRef]
18. Watanabe, R.; Minh-Dung, N.; Takahashi, H.; Takahata, T.; Matsumoto, K.; Shimoyama, I. Fusion of Cantilever and Diaphragm Pressure Sensors According to Frequency Characteristics. In Proceedings of the 2015 28th IEEE International Conference on Micro Electro Mechanical Systems (MEMS), Estoril, Portugal, 18–22 January 2015; pp. 212–214.
19. Wada, R.; Takahashi, H. Time Response Characteristics of a Highly Sensitive Barometric Pressure Change Sensor Based on MEMS Piezoresistive Cantilevers. *Jpn. J. Appl. Phys.* **2020**, *59*, 070906. [CrossRef]
20. Takahashi, H.; Dung, N.M.; Matsumoto, K.; Shimoyama, I. Differential Pressure Sensor Using a Piezoresistive Cantilever. *J. Micromech. Microeng.* **2012**, *22*, 055015. [CrossRef]
21. World Meteorological Organization. *Guide to Instruments and Methods of Observation (WMO-No. 8) 2021 Edition—Volume III: Observing Systems*; World Meteorological Organization: Geneva, Switzerland, 2021.
22. Lanzinger, E.; Schubotz, K. *A Laboratory Intercomparison of Static Pressure Heads*; World Meteorological Organization: Geneva, Switzerland, 2012; Volume 16.
23. National Oceanic and Atmospheric Administration; National Aeronautics and Space Administration; United States Air Force. *U.S. Standard Atmosphere, 1976*; U.S. Goverment Publishing Office: Washington, DC, USA, 1976.
24. Kundu, P.K.; Cohen, I.M.; Dowling, D.R. *Fluid Mechanics*, 5th ed.; Elsevier: Cambridge, MA, USA, 2012; ISBN 9780123821003.
25. Fox, R.W.; Pritchard, P.J.; McDonald, A.T. *Introduction to Fluid Mechanics*, 7th ed.; Wiley & Sons: Hoboken, NJ, USA, 2010.
26. Qorvo DWM1000—Qorvo. Available online: https://www.qorvo.com/products/p/DWM1000 (accessed on 1 August 2023).
27. TDSR UWB Module—Products—TDSR. Available online: https://tdsr-uwb.com/uwb-module/ (accessed on 1 August 2023).
28. Huang, Z.C.; Yeh, C.Y.; Tseng, K.H.; Hsu, W.Y. A UAV-RTK Lidar System for Wave and Tide Measurements in Coastal Zones. *J. Atmos. Ocean Technol.* **2018**, *35*, 1557–1570. [CrossRef]
29. Romero-Andrade, R.; Trejo-Soto, M.E.; Vázquez-Ontiveros, J.R.; Hernández-Andrade, D.; Cabanillas-Zavala, J.L. Sampling Rate Impact on Precise Point Positioning with a Low-Cost Gnss Receiver. *Appl. Sci.* **2021**, *11*, 7669. [CrossRef]
30. Marvelmind. Marvelmind Indoor Navigation System—Operating Manual. Available online: https://marvelmind.com/pics/marvelmind_navigation_system_manual.pdf (accessed on 14 October 2023).
31. Nguyen, A. MmWave Radar Sensors: Object Versus Range. Available online: https://www.ti.com/jp/lit/pdf/swra593 (accessed on 14 October 2023).

32. UV Verification Services Inc. RF Exposure Analysis for Millimeter Wave Radar Sensor Development Boards. Available online: https://dev.ti.com/tirex/explore/node?node=A__AIRQGZEx5ol-gjk4J66SuQ__radar_toolbox__1AslXXD__LATEST (accessed on 14 October 2023).
33. OMRON. 2SMPB-02B Digital Barometric Pressure Sensor. Available online: https://components.omron.com/us-en/asset/54931 (accessed on 14 October 2023).
34. Chapman, P.W.; Tufte, O.N.; Zook, J.D.; Long, D. Electrical Properties of Heavily Doped Silicon. *J. Appl. Phys.* **1963**, *34*, 3291–3295. [CrossRef]
35. Hagiwara, T.; Takahashi, H.; Takahata, T.; Shimoyama, I. Ground Effect Measurement of Butterfly Take-Off. In Proceedings of the 2018 IEEE Micro Electro Mechanical Systems (MEMS), Belfast, UK, 21–25 January 2018; pp. 832–835.

micromachines

MDPI

Article

Analysis of Acousto-Optic Phenomenon in SAW Acoustofluidic Chip and Its Application in Light Refocusing

Xianming Qin [1,2], Xuan Chen [3], Qiqi Yang [3], Lei Yang [3], Yan Liu [1,2], Chuanyu Zhang [3], Xueyong Wei [3,*] and Weidong Wang [1,2,*]

[1] School of Mechano-Electronic Engineering, Xidian University, Xi'an 710071, China; qinxianming@xidian.edu.cn (X.Q.)
[2] CityU-Xidian Joint Laboratory of Micro/Nano-Manufacturing, Xi'an 710071, China
[3] State Key Laboratory for Manufacturing Systems Engineering, Xi'an Jiaotong University, Xi'an 710049, China; chuanyu.zhang@xjtu.edu.cn (C.Z.)
* Correspondence: seanwei@mail.xjtu.edu.cn (X.W.); wangwd@mail.xidian.edu.cn (W.W.)

Abstract: This paper describes and analyzes a common acousto-optic phenomenon in surface acoustic wave (SAW) microfluidic chips and accomplishes some imaging experiments based on these analyses. This phenomenon in acoustofluidic chips includes the appearance of bright and dark stripes and image distortion. This article analyzes the three-dimensional acoustic pressure field and refractive index field distribution induced by focused acoustic fields and completes an analysis of the light path in an uneven refractive index medium. Based on the analysis of microfluidic devices, a SAW device based on a solid medium is further proposed. This MEMS SAW device can refocus the light beam and adjust the sharpness of the micrograph. The focal length can be controlled by changing the voltage. Moreover, the chip is also proven to be capable of forming a refractive index field in scattering media, such as tissue phantom and pig subcutaneous fat layer. This chip has the potential to be used as a planar microscale optical component that is easy to integrate and further optimize and provides a new concept about tunable imaging devices that can be attached directly to the skin or tissue.

Keywords: acoustofluidics; acousto-optic effect; SAW chip

Citation: Qin, X.; Chen, X.; Yang, Q.; Yang, L.; Liu, Y.; Zhang, C.; Wei, X.; Wang, W. Analysis of Acousto-Optic Phenomenon in SAW Acoustofluidic Chip and Its Application in Light Refocusing. *Micromachines* **2023**, *14*, 943. https://doi.org/10.3390/mi14050943

Academic Editor: Aiqun Liu

Received: 31 March 2023
Revised: 23 April 2023
Accepted: 25 April 2023
Published: 26 April 2023

1. Introduction

Surface acoustic wave (SAW) devices based on piezoelectric materials and interdigital electrodes (IDTs) have a long history and many applications, such as radars [1], filters [2], gratings [3], and acousto-optical modulators [4]. In recent years, SAW devices have been further applied to the field of acoustofluidics [5,6]. Based on microelectromechanical systems (MEMS) technology, SAW acoustofluidic devices can realize precise operation [7] on droplets, particles, and cells [8–14] in a noninvasive, label-free, and contactless manner, making it a powerful and advanced tool in fields such as biology [15], chemistry [16], and forensic analysis [17].

The technology of SAW acoustofluidic devices is gradually maturing, but there are still many phenomena that have not been explained. For example, in some acoustofluidic experiments, strange light and shade phenomena appear in the place where ultrasonic waves act [18]. Unlike acousto-optic modulators, these shades are caused by light perpendicular to the chip plane instead of parallel [19]. SAW devices are generally considered able to construct a stable standing field on the piezoelectric substrate plane [20], thus changing the refractive index of the waveguide to achieve an acousto-optic effect. However, in acoustofluidic devices, Rayleigh waves generated by SAW transducers leak into the fluid in the form of inhomogeneous waves. On the one hand, with complex acoustic pressure distribution in three dimensions, analysis of the acousto-optic effect in SAW acoustofluidic devices is more complicated, but on the other hand, it has the potential to realize different modulation performances based on SAW. However, instead of tunable lenses [21,22], SAW optical devices are mostly used as deflectors [23] or frequency shifters [24]. Unlike SAW devices, bulk wave (BW) acousto-optical devices can refocus or modulate light sources into different patterns or even clarify an image [25–28].

In this paper, we explain the acousto-optic phenomenon in SAW acoustofluidic devices with a light beam refraction simulation. Based on these analyses, a new kind of MEMS SAW acousto-optic device was developed. Modified from a focused surface acoustic wave (FSAW) microfluidic chip, this device can modulate the refractive index of the bonded medium and refocus the light beam perpendicular to its SAW plane to clarify the image. With its performance verified with scattering media, such as tissue phantom and porcine epithelium, this device has the potential to be integrated into microscale optical equipment and provides a new concept for in vivo acousto-optic imaging applications involving skin and other biological tissues.

2. Phenomenon and Analysis

2.1. Analysis of the Focused Acoustic Field

In SAW acoustofluidic devices, the position of the acousto-optic phenomenon is consistent with the position of the acoustic field at resonance [20,29,30]. Bright stripes are located in the center of the acoustic field along the propagation direction of the SAW beam, while darker stripes are distributed around it (Figure 1b–d). To analyze the acousto-optic phenomenon, we first simulated the three-dimensional distribution of the acoustic field in the fluid.

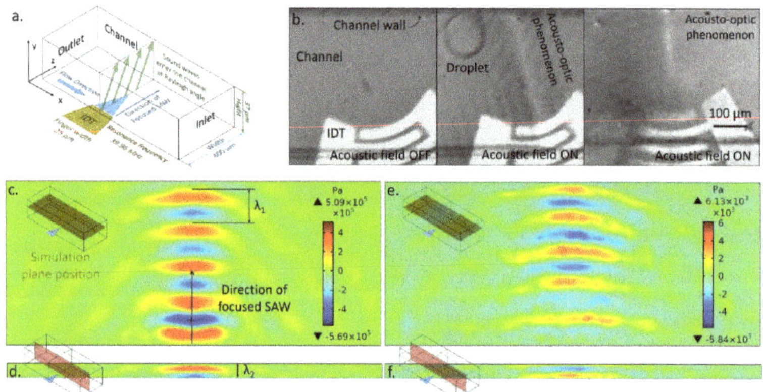

Figure 1. Analysis of the focused acoustic field: (**a**) Schematic diagram of the FSAW microfluidic device; (**b**) acousto-optic phenomena in SAW microfluidic devices; (**c,d**) acoustic pressure field distribution; (**e,f**) time-averaged acoustic pressure field distribution. The pink rectangle in the schematic diagram on the left of (**c–f**), respectively, refers to the plane where the simulation results are located in the channel.

In the COMSOL simulation of the acoustofluidics experiment, the substrate is 128-degree YX-cut lithium niobate. As shown in Table 1, the size of the substrate is 1500 μm × 1200 μm × 200 μm (length × width × thickness). The substrate thickness is set to 300 μm. The sound velocity on the surface of lithium niobate is 3996 m/s. The IDTs are modeled as perfect conductors using boundary conditions. Assuming the IDTs are very thin compared with the piezoelectric substrate, the effect of their mass and stiffness on the dynamics of the device is not accounted for. The model is solved at a target frequency of 39.96 MHz. The applied AC voltage is 35 V. The finger width and interspace of the IDTs are 25 μm. The acoustic window is 45°. The number of electrodes is set as 4 for computational efficiency. The simulation cannot calculate the coupling of the acoustic waves generated by all 50 electrodes, so it can only roughly show the distribution of the acoustic pressure field. The channel is

1000 µm × 37 µm × 475 µm (length × height × width). The channel is set to pure water at room temperature.

Table 1. COMSOL simulation parameters.

Parameter Name	Parameter Value
Designed frequency	39.96 MHz
Applied voltage	25 V
Finger width and interspace of the IDT	25 µm
Size of piezoelectric substrate	1500 µm × 1200 µm × 200 µm
Size of channel	1000 µm × 37 µm × 475 µm
Maximum calculation step time	1.25×10^{-9} s

The substrate is set as a piezoelectric material with a mechanical damping. The loss factor defined in the frequency domain at the target frequency is converted to the Rayleigh stiffness damping parameter. The bottom surface and all the surfaces on the sides of the substrate are assigned as the low-reflecting boundary condition. The microchannels are polydimethylsiloxane (PDMS) grooves processed with photolithography technology. The channel roof and channel walls are set with an acoustic impedance of 1070 kg/m^3 × 1030 m/s Pa·s/m, which reflects the PDMS boundary. In the simulation, the multiphysics interface "fluid–structure interaction" and "acoustic–structure boundary" are used to model the interaction between elastic waves in solids and pressure waves in fluids.

After a SAW enters the channel, it constructs a three-dimensional acoustic field in the liquid [31]. The wavelength λ1 of the SAW on the piezoelectric substrate is 100 µm, influenced by the IDT pattern, signal frequency, and properties of the substrate. When the sound wave enters the fluid medium inside the channel, the change in sound speed in the medium will lead to a change in wavelength. In the fluid, sound waves propagate in the form of an inhomogeneous wave with wavelength λ2, 37 µm. As shown in Figure 1, in the X-direction, i.e., the flow direction, the width of the acoustic pressure field obeys the size of the acoustic window of focusing IDT. In the Y-direction, i.e., the height direction, the high- and low-pressure regions are alternately distributed, with λ2 as the period. In the Z-direction, that is, in the SAW propagation direction and the direction of the angular bisector of the IDT, the high- and low-pressure regions of the acoustic pressure field are alternately distributed with λ1 as the period. There is a time scale difference between the flow in the channel, MHz acoustic waves, and beam propagation. For flow analysis, the acoustic field is time-averaged, while for light analysis, the acoustic field is static. Therefore, the transient simulation analysis can reflect the acoustic field through which the light passes, while the time-average acoustic field simulation can reflect the actual acoustic field distribution acting on the microfluidic targets. In Figure 1e,f, the time-averaged result is the distribution of the average acoustic pressure in ten periods. In the microfluidics experiments in Figure 1, the fluid is deionized water. The flow rate is regulated by a pressure pump (LSP01-2A, LongerPump, CN).

2.2. The Analysis of the Phenomenon through Simulation

The density of the fluid medium is modulated by the pressure waves generated from the SAW. The medium in the high-pressure region is compressed, leading to an increase in density, whereas, in the negative-pressure region, the opposite is true. The refractive index will change with the density. Therefore, the fluctuation of the acoustic pressure field will change the local refractive index of the medium as follows [32]:

$$n^2 = 1 + \sum_{i=1}^{3} \frac{(\alpha_i P^2 + \beta_i P + \delta_i)\lambda_L^2}{\lambda_L^2 - (A_i P^2 + B_i P + \Delta_i)} \tag{1}$$

where α_i, β_i, and δ_i and the corresponding Greek capital letters A_i, B_i, and Δ_i are the coefficients of the polynomials describing the Sellmeier dispersion coefficients considered pressure-dependent. n is the refractive index, and λ_L is the wavelength of the light. The study of Maysamreza Chamanzar [27] simplifies the relationship between the refractive index and pressure. With the fixed wavelength and temperature, the refractive index distribution can be calculated from the simulated pressure curve by using the following linear relationship:

$$n_{local} = n_0 + kP_{local} \tag{2}$$

where n_{local} is the local refractive index, n_0 is the refractive index of the medium at room temperature (25 °C) in the visible optical wavelength range, P_{local} is the local pressure, and k is an empirical coefficient. For water at room temperature, k is 1.402×10^{-5} bar^{-1} [33].

Based on the pressure distribution and the simplified relationship between the refractive index and pressure, the distribution of the refractive index can be calculated with the Runge Kutta method in MATLAB [34]. The calculation is performed on the X–Y planes, perpendicular to the propagation direction of SAW (Z-direction). The position of the plane is determined by the distance d_p from the plane to the channel wall.

Firstly, we model the refractive index field in MATLAB according to Equation (2). Since the relationship is linear (Equation (2)), the refractive index distribution simulated in MATLAB is very similar to the acoustic pressure field in the COMSOL simulation. To balance the accuracy and computation time, the simulation plane is divided into a 300 (X-direction) × 100 (Y-direction) grid, and all data points on the grid are initialized. The actual size of a single grid is about 3.3 × 0.37 μm. The Barron interpolation method is used to convert discrete index data points into continuous index fields on the grid. The refractive index gradient values of all data points in the X and Y directions are calculated, and a uniformly grided refractive index gradient field data is modeled.

Secondly, the light path is simulated in MATLAB. Light rays are set as incident perpendicular to the bottom of the channel. The step size of the Runge Kutta method is partially determined by the mesh size of the refractive index field. In addition, the step size varies according to the change in the refractive index. When the refractive index changes greatly, the step size will become smaller to avoid distortion. When the change in the refractive index exceeds 10^{-7}, 10^{-6}, 10^{-5}, and 10^{-4}, the step size will be reduced to 0.5, 0.2, 0.1, and 0.05 of the original step size, respectively. The number of rays is 321. There is an incident ray at x = 0 μm (that is, the location of the center line of the IDT), and 160 rays on its left and right side, respectively. The incident points of these 321 rays are uniformly distributed on the X-axis (bottom of the channel) and are within the range of x = (−500, 500) μm. For the light in a certain unit of the grid, the direction change is calculated based on the local refractive index gradient, and the optical path length is calculated based on the variable step size, so the displacement vector and position data are obtained. The Runge Kutta method is used to continuously calculate the position of the light rays until they reach the grid boundary.

Finally, the optical paths on multiple X–Y planes within the channel are simulated. The light enters vertically from the bottom of the channel, propagates in the liquid medium with a variable refractive index, and exits at the channel roof (Figure 2). To better understand the light path in the whole channel area, several vertical planes with different d_p are selected, and every light beam on each plane is simulated. The change in the X-coordinate from the incident position to the exit position of each light path is calculated (Figure 2d,e). It can be seen that in the two neighborhoods around the middle (x = 0), there is a peak on the left and a trough on the right, while the part near the center quickly approaches zero. This shows that in this variable refractive index field, the light on the left and right sides tends to approach the middle. The original uniformly distributed light converges to the center under the influence of the variable refractive index. In addition, in the area far away from the acoustic line, that is, in the range of about (−500, −100) μm and (100, 500) μm, the light path also has some small variation. In these areas, the pressure in the fluid is not constant, but the amplitude of the acoustic pressure is low, so the change in the refractive index is small. Although the X-coordinate variation in light is no more than 2 μm in the simulation, it is only the result of the change in light inside the microfluidic channel. The position change of the X-coordinate of light shows the change in the light's direction. The outgoing light from the top of the channel will go through a long light path till it forms a micro photo in the camera, which makes the original small change become an obvious acousto-optic phenomenon.

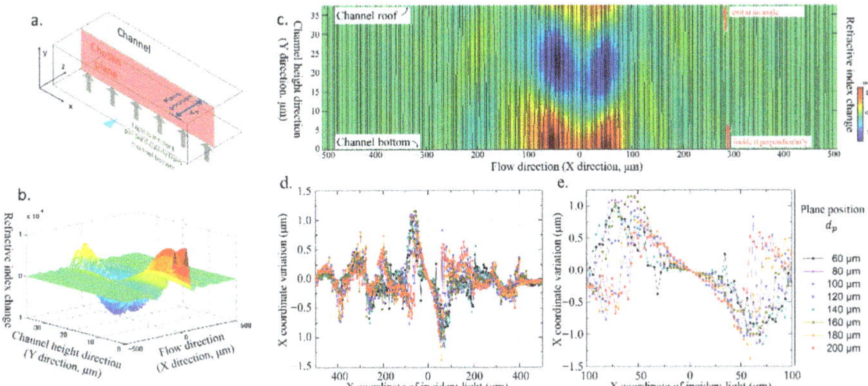

Figure 2. Simulation of refractive index distribution and optical path: (**a**) the X–Y plane selected for simulation, perpendicular to the bottom of the channel and the propagation direction of SAW; (**b**) the refractive index distribution on the selected plane; (**c**) the simulation result of 321 light paths on the selected plane; (**d**) the displacement of the light path on the X-axis; (**e**) a partial close-up of (**d**), the 200 μm wide central area around x = 0.

3. Application Experiments

To utilize and verify the analysis demonstrated above, we propose a SAW chip that can adjust focus. Manufactured with the MEMS process, this chip retains the design of focused IDT but does not retain any microchannels; instead, a 5 mm thick PDMS layer is used as the medium to conduct sound waves and light waves. In this way, the robustness of the device can be increased, and more importantly, pressure pumps or other microfluidics equipment are no longer needed to control the liquid. Under the action of a focused acoustic field, the acousto-optic phenomenon produces different patterns under different frequencies (Figure 3). From 38.0 MHz to 39.0 MHz, the change in vibration modes on the piezoelectric substrate influences the distribution of the refractive index inside PDMS. The voltage applied is 20 V.

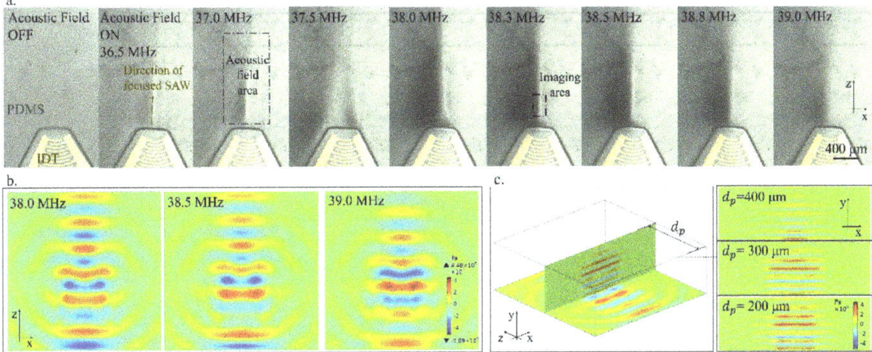

Figure 3. The SAW chip with PDMS as solid media: (**a**) acousto-optic phenomenon patterns under different frequencies; (**b**) the acoustic pressure distribution in X–Z plane under different frequencies; (**c**) the acoustic pressure field in the X–Y planes with different d_p.

Both the finger width and interspace of the IDT are 25 μm. The designed resonant frequency of the IDT transducer on the piezoelectric substrate is 39.96 MHz, but after bonding with PDMS, the actual resonant frequency will change to around 38.3 MHz. The acoustic window is 45°, and the geometric focal length is 200 μm. The amount of electrode finger pairs will affect the acoustic energy density and resonant frequency bandwidth. In order to form resonance, the number of finger pairs should not be more than 100. On this chip, each transducer has 50 pairs of electrodes. The chip manufacturing method and experimental equipment are the same as previously published articles [30].

For SAW chips with solid media in Figure 3, its acoustic pressure field is simulated in COMSOL (Figure 3b,c). The distribution of the acoustic pressure field in fluids (Figure 1) and elastic polymers (Figure 3) exhibits high similarity. In the X–Z plane, the focused acoustic wave beam (Figure 3b) determines the acoustic field area (Figure 3a), while in the X–Y plane, the propagation of acoustic waves generates a pressure field (Figure 3c) that modulates the refractive index field. Due to the situation that one of the most critical parameters, the Gladstone–Dale coefficient of PDMS, has not been reported before, the specific refractive index change in PDMS under the acousto-optic effect can hardly be calculated by theoretical simulation. However, according to the experimental results, it can be observed that the SAW chip with PDMS as the medium can effectively excite acousto-optic phenomena. As shown in Figure 3, the acousto-optic phenomenon exhibits different patterns (Figure 3a) under the acoustic fields at different frequencies (Figure 3b). In Figure 3a, the wavelength range of the light source is 400–760 nm, and the direction of the light is perpendicular to the X–Y plane. In the COMSOL simulation of Figure 3b,c, the modeling method is similar to the simulation in Figure 1, but the fluid medium with a $600 \times 200 \times 600$ μm (length × height × width) PDMS cube is replaced, and the acoustic impedance boundary condition is replaced with a low reflection boundary.

We used a very simple and straightforward method to demonstrate the MEMS SAW device's refocus ability. The chip is put on an opaque mask with transparent characters, through which the light source can form luminous characters (Figure 4). A microscope with a focal length of 45 mm is used to observe the characters' image and the modulation effect. First, the mask is put on the focal point, and the image is clear. Then, in order to imitate a blurred image, the mask is deliberately moved a certain distance away from the objective lens along the Y-direction to cause the image of the luminous characters to be out of focus. Finally, the acoustic field is turned on, and the voltage is adjusted until the light ray refocuses. In Figure 4b, the distance between the mask and the focal point is 1200 μm. As the voltage increases, the letter A in the image first gradually becomes recognizable and then blurs again, which demonstrates the process of equivalent focal length changing with an increase in acoustic field intensity. The size of the imaging area is subject to the region of the acoustic field. In the X-direction, the width of the imaging area is determined by the acoustic aperture. In the Z-direction, the length of the imaging area depends on the acoustic waves' attenuation length.

The equivalent focal length can be adjusted by changing the voltage to control the intensity of the acoustic field (Figure 5). We used chips with the same design in the experiments to test their refocus effect. The relationship between the equivalent focal length variation and the voltage is approximately linear (Figure 5b). A linear relation exists between the applied voltage of the transducer responsible for exciting the ultrasonic resonance and the induced acoustic pressure amplitude [35], which results in a linear relationship between the voltage and the local refractive index, according to Equation (2). Each data point is the average of four measurements. The device has a settling time when the acoustic field is turned on. This settling time is not decided by the sound velocity. The acoustic field needs a certain amount of time to accumulate before the amplitude of the acoustic pressure field reaches the maximum and stability. Similarly, when the power is turned off, it also takes a certain period of time for the acoustic field to gradually decline before it disappears due to the acoustic waves' attenuation. With an increase in voltage, no matter whether the power is on or off, the settling time will be prolonged (Figure 5c). There is a linear relationship between acoustic energy density and the square of voltage [35], which results in a direct proportional relationship between the settling time and voltage. Each data point is the average of five tested settling times.

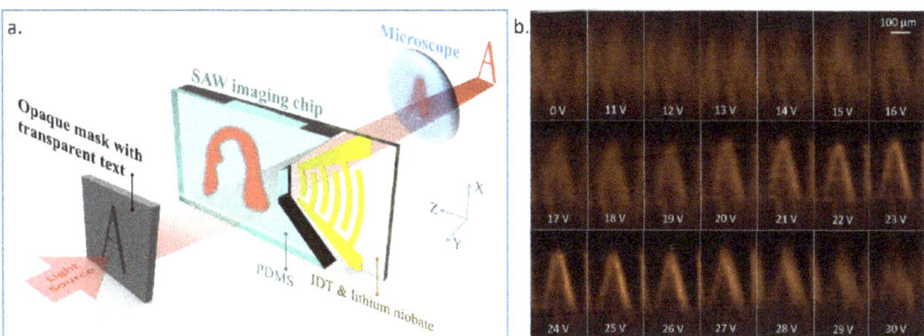

Figure 4. The method of testing the new SAW chip and its performance. (**a**) Schematic diagram of device structure and imaging method. (**b**) Images modulated by acoustic fields with different voltages.

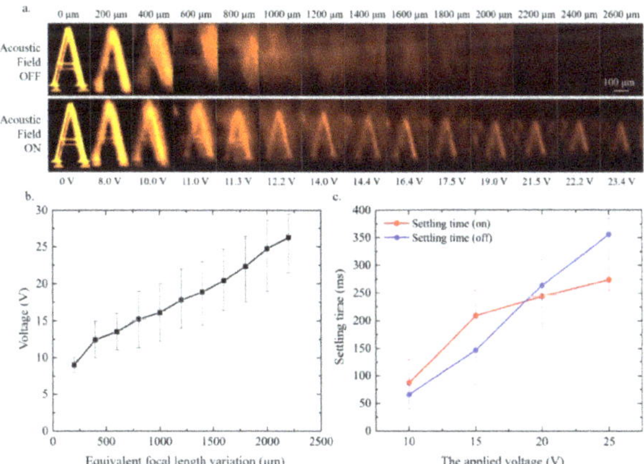

Figure 5. Relationship between voltage and device performance: (**a**) for every two images in the column, the data above are the distance between the mask and the focal point, and the data below are the voltage applied on the transducer; (**b**) the average result of the relationship between voltage and equivalent focal length variation; (**c**) the relationship between applied voltage and settling time. The settling time (on) and the settling time (off) refer to the time from power turned on or off to the stabilization of the image, respectively.

Because the MEMS SAW chip is based on the planar structure of a lithium niobate (LN) wafer, it can be put in different positions or on different media to modulate the image. Firstly, we tested the chip's performance for dynamic images. By moving the chip and mask in the opposite direction, the light in different regions is refocused (Figure 6a,b). The voltage applied to the device is 24.7 V, and the distance between the mask and the focal point is 1600 μm. Since the PDMS waveguide bonded to the transducer is flexible, the chip has a good conformal ability, which makes it adaptable on uneven surfaces. The video of the dynamic image modulation is in the Supplementary Materials.

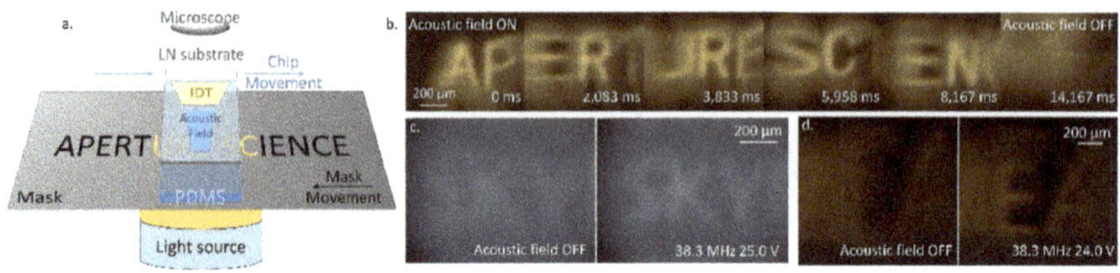

Figure 6. Experiments with dynamic image and scattering medium: (**a**) schematic diagram of chip position and movement; (**b**) experimental results of dynamic image modulation; (**c**) experiments with tissue phantom as scattering medium; (**d**) experiments with pig subcutaneous fat layer as scattering medium.

Secondly, we tested the chip's performance on different materials. Scattering media, such as human skin and tissue, can disperse the light beams passing through, making the captured image blurred. Ultrasonic waves undergo minimal attenuation (about $0.3{\sim}0.6$ dB cm^{-1} MHz^{-1}) when propagating through biological tissue [36,37], which makes it possible for the SAW chip to refocus the light beam inside skin or tissue by pasting the transducer directly on them. In the experiment, we used titanium dioxide mixed with PDMS as tissue phantom, which can simulate the human epithelium's optical properties and has tunable reduced scattering and absorption coefficients at visible and near-infrared wavelengths [36,37]. Although the scattering coefficient and absorption coefficient of tissue phantom (TiO$_2$/PDMS) are different from transparent medium (pure PDMS), their similarity in acoustic characteristics makes it possible for SAW to modulate the refractive index distribution and sharpen the image (Figure 6c,d). For the chip with tissue phantom, the concentration of titanium dioxide is 3 ‰, and the reduced scattering coefficient and absorption coefficient are about 5 cm^{-1} and 0.03 cm^{-1}, respectively. The thickness of the tissue phantom is about 2 mm. As a solid medium and a replacement for PDMS, tissue phantom is directly bonded to the piezoelectric substrate using oxygen plasma bombardment.

In addition to tissue phantom, some biological tissues were used for similar experiments (Figure 6d). The subcutaneous fat layer of pigskin (Oxygen Crown, CN) is cut into about a 1 mm thick layer and placed between the device and the mask as a scattering medium. Usually, the characters can hardly be recognized under the pig fat layer. After the acoustic field is turned on, some characters can be vaguely distinguished. However, it should be noted that the fat layer is only pressed on the chip instead of bonded. If the chip is tightly bonded to the adipose layer or other similar tissue so that sound waves can construct a better index field inside, the imaging quality will be better.

4. Discussion

The principle of the acousto-optic effect in acousto-optical modulators and SAW microfluidic devices is the same, but the relative position of light and acoustic field is different, resulting in different acousto-optic phenomena. In SAW acousto-optic modulators, such as tunable SAW optical switches, light propagates along the surface plane of the piezoelectric substrate. For SAW microfluidic devices, the beam is perpendicular to the SAW plane and travels in the three-dimensional acoustic pressure field, resulting in different performances of light modulation. Based on the analysis of the three-dimensional refractive index field and light path, a new kind of SAW imaging device was developed, and its performance was tested through a series of experiments.

This MEMS SAW device can extend the focal length of the microscope within the range of 0~2000 µm. Although within the range of 2000~2600 µm, it can still achieve light refocusing, but the image is difficult to recognize due to its small size. Although the device cannot directly change the resolution of an image, the sharpness of the image can be adjusted by controlling the voltage. The imaging speed of this device is influenced by voltage, and the image settling time varies from 50 to 300 ms in a voltage range of 10~30 V. The size of the imaging area of this device is controlled by an acoustic aperture width in the X-direction and an acoustic field in the Z-direction, which is approximately 200 µm (X-direction) × 600 µm (Z-direction).

Compared with a tunable lens [25], this surface acoustic wave device aims at an in vivo imaging device. Although it has a zooming effect, its purpose is not to replace the lens but to change the refractive index inside the observed object and turn the target into an equivalent convex lens.

Compared with a cylindrical acousto-optic relay lens [26], the acousto-optic interaction distance of this SAW chip is short. The acousto-optic interaction distance of a cylindrical acoustic relay lens can reach tens of millimeters [27], while the interaction distance in our SAW chip depends on the attenuation of the acoustic field, that is, the height of the acoustic field in the Y-direction. Its short acousto-optic interaction distance causes the maximum acousto-optic modulation effect of our chip to be weaker than cylindrical ones as a sacrifice to balance its size and performance. However, the cylindrical acousto-optic relay lens is based on hollow cylindrical acoustic transducer arrays, so the target sample needs to be wrapped in a hollow cylinder to achieve ultrasonic sculpting of virtual optical waveguides, which limits the size and type of the sample to be observed [28]. For example, a cylindrical ultrasonic transducer cannot be used for samples larger than the cylinder. In addition, the planar structure of this device allows it to be attached directly to the target sample's surface, allowing it to be further optimized to work as a planar microscale optical component that is easy to integrate.

5. Conclusions

The acousto-optic phenomena in acoustic microfluidic devices are quite different from those in acousto-optic modulators. In this paper, we analyzed the refractive index field that an acoustic wave constructs in a medium and explained the phenomenon. Based on the analysis, a new kind of SAW chip was developed that can refocus the light to make an originally blurred image clear. Through a series of experiments, the relationship between the voltage and equivalent focal length variation was established. Experiments with the dynamic image and tissue phantom showed that this device has the potential to be used in in vivo imaging, such as observing fluorescently labeled tissues that are blocked by scattering media.

Supplementary Materials: The following supporting information can be downloaded at: https://www.mdpi.com/article/10.3390/mi14050943/s1, Video S1.

Author Contributions: Conceptualization, X.Q.; investigation, X.Q.; methodology, X.Q.; visualization, X.Q.; writing—original draft, X.Q.; writing—review & editing, X.C., Q.Y., L.Y. and Y.L.; Supervision, C.Z., X.W. and W.W.; Resources, X.W. All authors have read and agreed to the published version of the manuscript.

Funding: This work is sponsored by the National Key Research and Development Program (Grant No. 2022YFB3204800), the Key Research and Development Program of Shaanxi (Program No. 2022NY-211), the Youth Innovation Team of Shaanxi Universities (Grant No. 2022-63), and the Fundamental Research Funds for the Central Universities (Grant No. XJSJ23123). We also appreciate the support of the International Joint Laboratory for MicroNano Manufacturing and Measurement Technologies.

Data Availability Statement: Not applicable.

Conflicts of Interest: The authors declare no conflict of interest.

References

1. Dransfeld, K.; Salzmann, E. Excitation, detection, and attenuation of high-frequency elastic surface waves. *Phys. Acoust.* **2012**, *7*, 219–272.
2. Ruppel, C.C.W. Acoustic wave filter technology—A review. *IEEE Trans. Ultrason. Ferroelectr. Freq. Control* **2017**, *64*, 1390–1400. [CrossRef]
3. Li, R.C.M.; Melngailis, J. The influence of stored energy at step discontinuities on the behaviour of surface-wave gratings. *IEEE Trans.* **1975**, *22*, 189–198.

4. Dühring, M.B.; Sigmund, O. Improving the acousto-optical interaction in a Mach–Zehnder interferometer. *J. Appl. Phys.* **2009**, *105*, 083529. [CrossRef]

5. Fogel, R.; Limson, J.; Seshia, A.A. Acoustic biosensors. *Essays Biochem.* **2016**, *60*, 101–110.

6. Shi, J.; Ahmed, D.; Mao, X.; Lin, S.-C.S.; Lawit, A.; Huang, T.J. Acoustic tweezers: Patterning cells and microparticles using standing surface acoustic waves (SSAW). *Lab Chip* **2009**, *9*, 2890–2895. [CrossRef]

7. Bruus, H. Acoustofluidics 1: Governing equations in microfluidics. *Lab Chip* **2011**, *11*, 3742–3751. [CrossRef] [PubMed]

8. Ozcelik, A.; Rufo, J.; Guo, F.; Gu, Y.; Li, P.; Lata, J.; Huang, T.J. Acoustic tweezers for the life sciences. *Nat. Methods* **2018**, *15*, 1021–1028. [CrossRef]

9. Peng, X.; He, W.; Xin, F.; Genin, G.M.; Lu, T.J. Standing surface acoustic waves, and the mechanics of acoustic tweezer manipulation of eukaryotic cells. *J. Mech. Phys. Solids* **2020**, *145*, 104134. [CrossRef]

10. Meng, L.; Cai, F.; Li, F.; Zhou, W.; Niu, L.; Zheng, H. Acoustic tweezers. *J. Phys. D Appl. Phys.* **2019**, *52*, 273001. [CrossRef]

11. Tian, Z.; Yang, S.; Huang, P.-H.; Wang, Z.; Zhang, P.; Gu, Y.; Bachman, H.; Chen, C.; Wu, M.; Xie, Y.; et al. Wave number–spiral acoustic tweezers for dynamic and reconfigurable manipulation of particles and cells. *Sci. Adv.* **2019**, *5*, eaau6062. [CrossRef] [PubMed]

12. Ding, X.; Peng, Z.; Lin, S.C.S.; Huang, T.J. Cell separation using tilted-angle standing surface acoustic waves. *Proc. Natl. Acad. Sci. USA* **2014**, *111*, 12992–12997. [CrossRef]

13. Wang, Y.; Pan, H.; Mei, D.; Xu, C.; Weng, W. Programmable motion control and trajectory manipulation of microparticles through tri-directional symmetrical acoustic tweezers. *Lab Chip* **2022**, *22*, 1149–1161. [CrossRef] [PubMed]

14. Go, D.B.; Atashbar, M.; Ramshani, Z.; Chang, H.-C. Surface acoustic wave devices for chemical sensing and microfluidics: A review and perspective. *Anal. Methods* **2017**, *9*, 4112–4134. [CrossRef]

15. Bin Mazalan, M.; Noor, A.M.; Wahab, Y.; Yahud, S.; Zaman, W.S.W.K. Current development in interdigital transducer (IDT) surface acoustic wave devices for live cell in vitro studies: A review. *Micromachines* **2021**, *13*, 30. [CrossRef] [PubMed]

16. Alshehhi, F.; Waheed, W.; Al-Ali, A.; Abu-Nada, E.; Alazzam, A. Numerical Modeling Using Immersed Boundary-Lattice Boltzmann Method and Experiments for Particle Manipulation under Standing Surface Acoustic Waves. *Micromachines* **2023**, *14*, 366. [CrossRef] [PubMed]

17. Norris, J.V.; Evander, M.; Horsman-Hall, K.M.; Nilsson, J.; Laurell, T.; Landers, J.P. Acoustic Differential Extraction for Forensic Analysis of Sexual Assault Evidence. *Anal. Chem.* **2009**, *81*, 6089–6095. [CrossRef]

18. Jung, J.H.; Destgeer, G.; Ha, B.; Park, J.; Sung, H.J. On-demand droplet splitting using surface acoustic waves. *Lab Chip* **2016**, *16*, 3235–3243. [CrossRef]

19. Kakio, S. Acousto-optic modulator driven by surface acoustic waves. *Acta Phys. Pol. A* **2015**, *127*, 15–19. [CrossRef]

20. Cai, L.; Mahmoud, A.; Khan, M.; Mahmoud, M.; Mukherjee, T.; Bain, J.; Piazza, G. Acousto-optical modulation of thin film lithium niobate waveguide devices. *Photon. Res.* **2019**, *7*, 1003–1013. [CrossRef]

21. Nagelberg, S.; Zarzar, L.D.; Subramanian, K.; Sresht, V.; Blankschtein, D.; Barbastathis, G.; Kreysing, M.; Swager, T.M.; Kolle, M. Reconfigurable and dynamically tunable droplet-based compound micro-lenses. In Proceedings of the 3D Image Acquisition and Display: Technology, Perception and Applications, Orlando, FL, USA, 25–28 June 2018; p. 3W3G.2.

22. Arbabi, E.; Arbabi, A.; Kamali, S.M.; Horie, Y.; Faraji-Dana, M.; Faraon, A. MEMS-tunable dielectric metasurface lens. *Nat. Commun.* **2018**, *9*, 812. [CrossRef] [PubMed]

23. Zuwei, Z.; Zhiyu, W.; Jing, H. Theoretical analysis and concept demonstration of a novel MOEMS accelerometer based on Raman—Nath diffraction. *J. Semicond.* **2012**, *33*, 044005.

24. Shao, L.; Yu, M.; Maity, S.; Sinclair, N.; Zheng, L.; Chia, C.; Shams-Ansari, A.; Wang, C.; Zhang, M.; Lai, K.; et al. Microwave-to-optical conversion using lithium niobate thin-film acoustic resonators. *Optica* **2019**, *6*, 1498–1505. [CrossRef]

25. Higginson, K.A.; Costolo, M.A.; Rietman, E.A. Adaptive geometric optics derived from nonlinear acoustic effects. *Appl. Phys. Lett.* **2004**, *84*, 843–845. [CrossRef]

26. Karimi, Y.; Scopelliti, M.G.; Do, N.; Alam, M.-R.; Chamanzar, M. In situ 3D reconfigurable ultrasonically sculpted optical beam paths. *Opt. Express* **2019**, *27*, 7249–7265. [CrossRef]

27. Scopelliti, M.G.; Chamanzar, M. Ultrasonically sculpted virtual relay lens for in situ microimaging. *Light Sci. Appl.* **2019**, *8*, 65. [CrossRef] [PubMed]

28. Chamanzar, M.; Scopelliti, M.G.; Bloch, J.; Do, N.; Huh, M.; Seo, D.; Iafrati, J.; Sohal, V.S.; Alam, M.-R.; Maharbiz, M.M. Ultrasonic sculpting of virtual optical waveguides in tissue. *Nat. Commun.* **2019**, *10*, 92. [CrossRef]

29. Qin, X.; Wang, H.; Wei, X. Intra-droplet particle enrichment in a focused acoustic field. *RSC Adv.* **2020**, *10*, 11565–11572. [CrossRef]

30. Qin, X.; Wei, X.; Li, L.; Wang, H.; Jiang, Z.; Sun, D. Acoustic valves in microfluidic channels for droplet manipulation. *Lab Chip* **2021**, *21*, 3165–3173. [CrossRef] [PubMed]

31. Shi, J.; Yazdi, S.; Lin, S.C.S.; Ding, X.; Chiang, I.-K.; Sharp, K.; Huang, T.J. Three-dimensional continuous particle focusing in a microfluidic channel via standing surface acoustic waves (SSAW). *Lab Chip* **2011**, *11*, 2319–2324. [CrossRef] [PubMed]

32. Weiss, L.; Tazibt, A.; Tidu, A.; Aillerie, M. Water density and polarizability deduced from the refractive index determined by interferometric measurements up to 250 MPa. *J. Chem. Phys.* **2012**, *136*, 124201. [CrossRef] [PubMed]

33. Waxler, R.; Weir, C.E. Effect of pressure and temperature on the refractive indices of benzene, carbon tetrachloride, and water. *J. Res. Natl. Bur. Stand. Sect. A Phys. Chem.* **1963**, *67A*, 163–171. [CrossRef] [PubMed]

34. Feng, D.H.; Pan, S.; Wang, W.L.; Li, H. Simulation and Analysis of Ray Tracing in Discretionary Gradient Refraction Index Medium. *Comput. Simul.* **2010**, *27*, 135–139.
35. Bruus, H. Acoustofluidics 7: The acoustic radiation force on small particles. *Lab Chip* **2012**, *12*, 1014–1021. [CrossRef] [PubMed]
36. Greening, G.J.; Istfan, R.; Higgins, L.M.; Balachandran, K.; Roblyer, D.; Pierce, M.C.; Muldoon, T.J. Characterization of thin poly(dimethylsiloxane)-based tissue-simulating phantoms with tunable reduced scattering and absorption coefficients at visible and near-infrared wavelengths. *J. Biomed. Opt.* **2014**, *19*, 115002. [CrossRef] [PubMed]
37. Hoskins, P.R.; Martin, K.; Thrush, A. *Diagnostic Ultrasound: Physics and Equipment*; CRC Press: Boca Raton, FL, USA, 2019.

 *micromachines*

Article

Flexible Thermoelectric Type Temperature Sensors Based on Graphene Fibers

Chenying Wang [1,2,3,†], Yuxin Zhang [4,†], Feng Han [1,2,3,*] and Zhuangde Jiang [2,3]

[1] School of Instrument Science and Technology, Xi'an Jiaotong University, Xi'an 710049, China; wangchenying@mail.xjtu.edu.cn
[2] State Key Laboratory for Manufacturing Systems Engineering, Xi'an Jiaotong University, Xi'an 710049, China; zdjiang@xjtu.edu.cn
[3] International Joint Laboratory for Micro/Nano Manufacturing and Measurement Technologies, Xi'an Jiaotong University, Xi'an 710049, China
[4] The 41st Institute of the Fourth Academy of CASC, Xi'an 710049, China; yxz_2023@163.com
[*] Correspondence: hanfeng_20625@xjtu.edu.cn
[†] These authors contributed equally to this work.

Abstract: Graphene, as a novel thermoelectric (TE) material, has received growing attention because of its unique microstructure and excellent thermoelectric properties. In this paper, graphene fibers (GFs) are synthesized by a facile microfluidic spinning technique using a green reducing agent (vitamin C). The GFs have the merits of high electrical conductivity (2448 S/m), high flexibility, and light weight. Further, a flexible temperature sensor based on GF and platinum (Pt) with a sensitivity of 29.9 μV/°C is proposed, and the thermal voltage output of the sensor can reach 3.45 mV at a temperature gradient of 120 °C. The sensor has good scalability in length, and its sensitivity can increase with the number of p-n thermocouples. It has good cyclic stability, repeatability, resistance to bending interference, and stability, showing great promise for applications in real-time detection of human body temperature.

Keywords: graphene fibers; vitamin C; thermoelectric; flexible temperature sensors

Citation: Wang, C.; Zhang, Y.; Han, F.; Jiang, Z. Flexible Thermoelectric Type Temperature Sensors Based on Graphene Fibers. *Micromachines* **2023**, *14*, 1853. https://doi.org/10.3390/mi14101853

Academic Editors: Weidong Wang, Yong Ruan and Zai-Fa Zhou

Received: 27 August 2023
Revised: 22 September 2023
Accepted: 25 September 2023
Published: 28 September 2023

1. Introduction

Changes in human body temperature are one of the most important indicators of human activity and health, helping to maintain the thermal balance between the body and its surroundings, while some physical sensations related to emotions can also be observed by analyzing subtle changes in body temperature. It is also known that changes in human body temperature can be used as a primary medical diagnostic assessment indicator for new crown pneumonia. Flexible wearable temperature sensors provide a method for real-time monitoring of human body temperature, which can be classified into thermal resistance, thermocouple, and fiber optic types according to the temperature measurement principle [1–3]. Among them, thermocouple temperature sensors are usually made of two materials. Based on the thermoelectric principle of temperature measurement, the temperature signal is converted into a voltage signal for output, and the resulting thermoelectric potential has a linear or near-linear relationship with the temperature difference [4]. Traditional thermocouple temperature sensors usually use metal as the thermal electrode, which is rigid and difficult to bend and cannot accurately measure the temperature on complex curved surfaces [1]. In addition, the high cost of precious metal materials limits the further development of thermocouple temperature sensors. With the increasing demand for comfort in wearable devices, flexible thermocouple temperature sensors, which are light in quality and highly flexible, are gradually gaining popularity. Benefiting from its flexibility and adaptability, it can be directly attached to the human skin to record the temperature of the human body and the external environment for a long time. Thermoelectric materials are one of the most important components in thermocouple temperature sensor research. Compared with traditional thermoelectric materials, flexible thermoelectric materials are bendable in shape, small in size, and light in mass and have good application prospects in wearable devices and other flexible electronics. Graphene is a thermoelectric material with excellent performance that has the advantages of non-toxicity, abundant raw materials, high conductivity, high electron mobility and carrier concentration, etc. [5–7]. Dragoman et al. proposed a graphene-based metal electrode device, according to which the transfer matrix method was used to predict the Seebeck coefficients up to 30 mV/K [8].

Graphene fiber (GF) is a type of macrofiber material that consists of individual units of graphene and its derivatives. There have been numerous methods employed to directly construct macrofibers using graphene sheets [9,10]. This material harnesses the remarkable characteristics of a single graphene sheet, including its exceptional mechanical strength, electrical conductivity, and thermal conductivity. Not only does it have mechanical flexibility and knittability similar to that of conventional fibers, but it is also lower cost, lighter, and more malleable than conventional carbon nanotube fibers, while it is easier to functionalize it in situ or in post-synthesis, which is of significant practical application value. However, the graphene fiber-based flexible thermocouples obtained from the current study still suffer from the problem of low sensitivity.

In this paper, graphene oxide fibers (GOFs) were proposed utilizing a microfluidic spinning technique, and graphene fibers were synthesized using vitamin C as a reducing agent. The impact of various concentrations of vitamin C in the graphene oxide (GO) solution on the electrical conductivity and thermoelectric properties of GF was examined. A lightweight, highly flexible, and inflexible thermocouple temperature sensor was designed using GF and platinum (Pt) electrodes and calibrated and tested with a sensitivity of up to 29.9 μV/°C. The sensor has good scalability in terms of length and can be made more sensitive as the number of thermocouple pairs comprising it increases.

2. Experimental Section

2.1. Materials and Reagents

The modified Hummer's method [11] was used to prepare the GO. Vitamin C was purchased from Sigma-Aldrich Co., Ltd., St. Louis, MO, USA.

2.2. Preparation of Graphene Fibers

An approach to fabricating graphene fibers using a facile microfluidic spinning method is proposed, as illustrated in Figure 1. Firstly, a well-dispersed 12 mg/mL graphene oxide solution was prepared after ultrasonication for 30 min. Vitamin C was added to the GO solution at different concentrations of 0 wt%, 1 wt%, 2 wt%, 3 wt%, and 4 wt%. These concentrations were labeled as GOFs, 1 wt%-GFs, 2 wt%-GFs, 3 wt%-GFs, and 4 wt%-GFs, respectively. The mixture was then thoroughly mixed to ensure homogeneity. Secondly, the graphene oxide solution was slowly injected into polytetrafluoroethylene (PTFE) tubes with a 0.5 mm inner diameter. The PTFE tube was put into an oven and subjected to a temperature of 150 °C for a duration of 2 h. While the water steadily evaporated, fibers began to develop within the tube. Subsequently, GOFs and GFs matching the geometry of the tube were pushed out of the tube with the help of N2 flow and dried in an atmospheric environment.

2.3. Fabrication of TE Temperature Sensors

As exhibited in Figure 2, the flexible temperature sensor was formed by combining two materials, with GF as the anode and the platinum (Pt) electrode as the cathode. The GF (about 2 cm long) and the Pt (about 2 cm long) were connected on a copper foil tape through the conductive silver paste and then mounted on a polyimide (PI) substrate (40 mm long × 20 mm wide). The connecting end of the two electrodes was the hot end of the thermoelectric-type temperature sensor, and the other ends of the anode and cathode electrodes were both the cold ends of the sensor. During the heating process, the hot end of the senor was put on the hot plate and heated by the silicone rubber heating belt as the heat source, while the cold end was placed in the air at 25 °C. The temperature of the hot plate can be controlled and adjusted by an intelligent temperature control system. The TE temperature sensor was obtained by applying a temperature difference between the hot and cold ends to measure the output voltage.

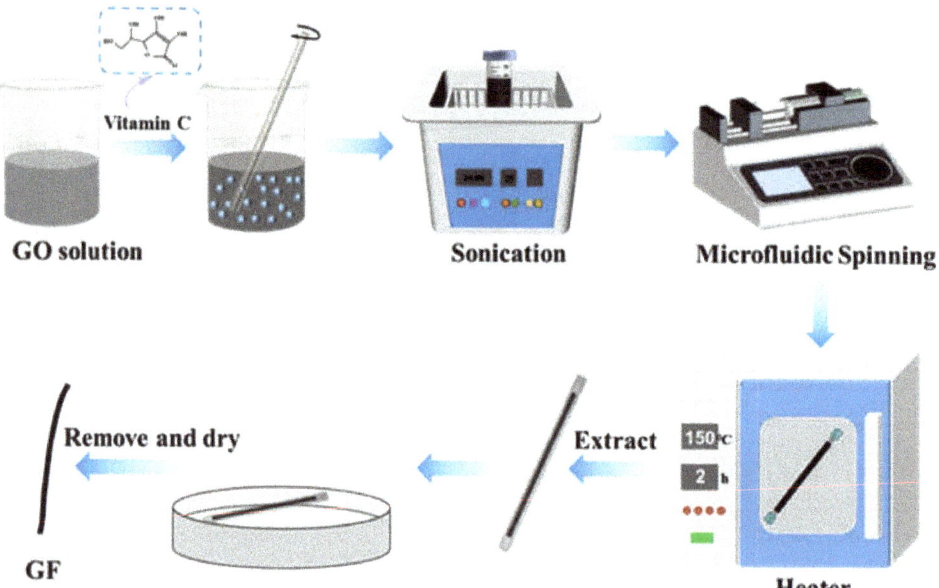

Figure 1. Schematic diagram of preparation process of GFs.

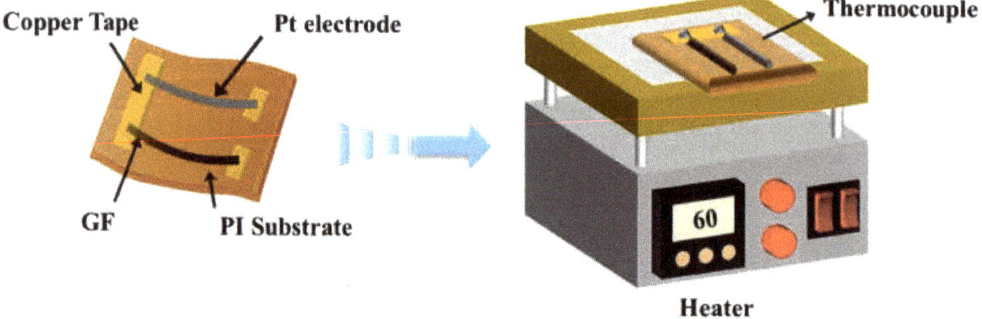

Figure 2. GF/Pt thermocouple preparation diagram.

2.4. Characterization and Measurements

A field emission scanning electron microscope (FESEM, SU-8010, Hitachi, Tokyo, Japan) was employed to capture morphological images of fibers. And a laser Raman spectrometer (HR800, Horiba Jobin Yvon, Paris, France) was applied to get the Raman spectroscopy. The chemical composition and electronic structure of the sample surface were analyzed using a Thermo Fisher Escalab xi + X-ray photoelectron spectroscopy analyzer (XPS, ESCALAB Xi+, Thermo Fisher, Waltham, MA, USA). Conductivities were measured using Source Meter (2450, Keithley, Beaverton, OR, USA). The output voltage was measured using a data collector (LR8450, HIOKI, Nagano, Japan). A thermoelectric performance test system was established that can calibrate and test the sensor in a wide low-temperature range. The hot end of the thermocouple was heated by the silicone rubber heating belt as the heat source, and its temperature was controlled and adjusted by an intelligent temperature control system. Additionally, the data collection device was utilized to measure the temperature at both the hot and cold ends, which were monitored using a standard K-type thermocouple. It also recorded the output voltage of the developed thermocouple.

The sensitivity (Seebeck coefficient) of a thermocouple temperature sensor is calculated using the following formula [12]:

$$S = \frac{dV}{d\Delta T} = \frac{dV}{d(T_h - T_c)} \tag{1}$$

V—Output thermopotential of the sensor; T_h—Temperature of the hot end; T_c—Temperature of the cold end; S—Sensitivity of the thermocouple.

3. Results and Discussion

The surface morphologies of the fibers with varying levels of vitamin C were investigated by SEM, as depicted in Figure 3. Figure 3a,b show the surface morphologies of GOF with a diameter of about 110 µm captured at different magnifications of 600 and 2 k, respectively. Compared with GOF, the diameter of GF is significantly reduced. GOF has a paper-like and less wrinkled surface with a disordered, smooth-surfaced fluffy appearance, whereas GF shows significant curvature and more wrinkles. The diameter of the fibers decreases gradually with the increase in vitamin C doping. The surface morphology of GF with different amounts of vitamin C doping shows no significant difference, and all show obvious bending and folded structures. The presence of van der Waals forces and the strong π-π stacking between graphene sheets cause GO sheets to easily clump together. However, the introduction of vitamin C prevents this clumping. Meanwhile, GO has abundant oxygen functional groups, including carboxyl (COOH), carbonyl (-C=O), epoxide (C-O-C), and hydroxyl (-OH) groups, whereas vitamin C can remove a large number of oxygen groups within GO during the chemical reduction process, leading to the appearance of wrinkles after the reduction process [13–15].

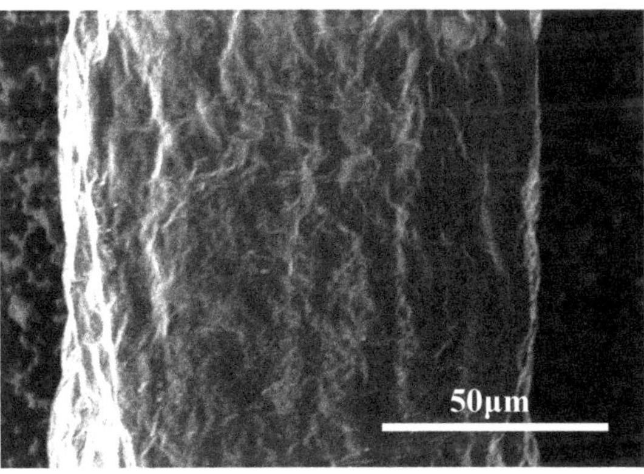

(**a**) GOF (600 times)   (**b**) GOF-2 (2 k times)

Figure 3. *Cont.*

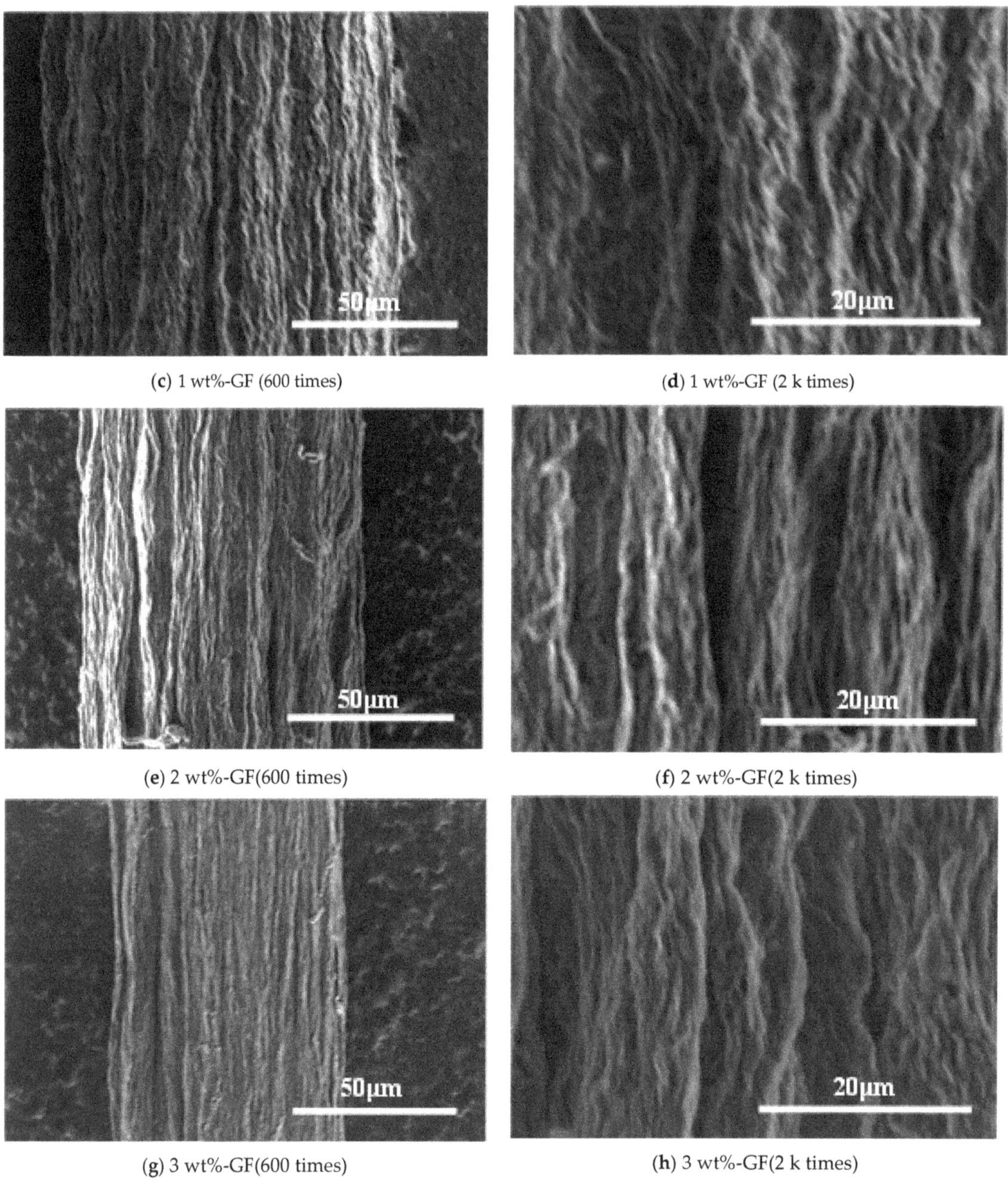

(**c**) 1 wt%-GF (600 times)

(**d**) 1 wt%-GF (2 k times)

(**e**) 2 wt%-GF(600 times)

(**f**) 2 wt%-GF(2 k times)

(**g**) 3 wt%-GF(600 times)

(**h**) 3 wt%-GF(2 k times)

Figure 3. *Cont.*

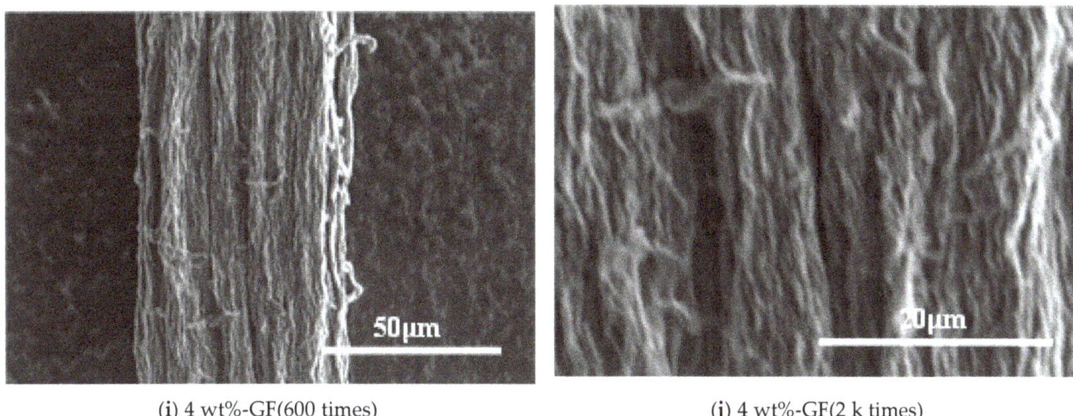

(**i**) 4 wt%-GF(600 times)   (**j**) 4 wt%-GF(2 k times)

Figure 3. SEM images of the GOFs and GFs with different vitamin C doping content.

The Raman spectra of GOFs and GFs with different C doping are illustrated in Figure 4. Raman spectra are widely used for the structural characterization of graphene and related materials. There are two distinct peaks observed in the spectrum. The first peak, known as the D peak, appears at 1350 cm^{-1} and is attributed to the telescopic vibration of the sp^3 carbon atom. This peak is indicative of structural deficiencies and disorders. The second peak, called the G peak, is located at 1590 cm^{-1} and is associated with the E$_{2g}$ vibrational mode of the sp^2 carbon domain [16,17]. The degree of defects and disorder in graphene structures can usually be characterized by the ratio of relative intensities of the D and G peaks (defined as I_D/I_G). After vitamin C reduction, the I_D/I_G values of 1 wt%-GF, 2 wt%-GF, 3 wt%-GF, and 4 wt%-GF are 1.02, 1.07, 1.08, and 1.11, respectively, which all increase compared with that of GOF (1.01). The increased I_D/I_G values indicate that the oxygen-containing functional groups were effectively removed during the chemical reduction process, leading to the introduction of more structural defects. Carbon-carbon bonds on the surface of the graphene structural layer are broken, and after reduction, the new forming sp^2 domains are smaller in size but more ubiquitous.

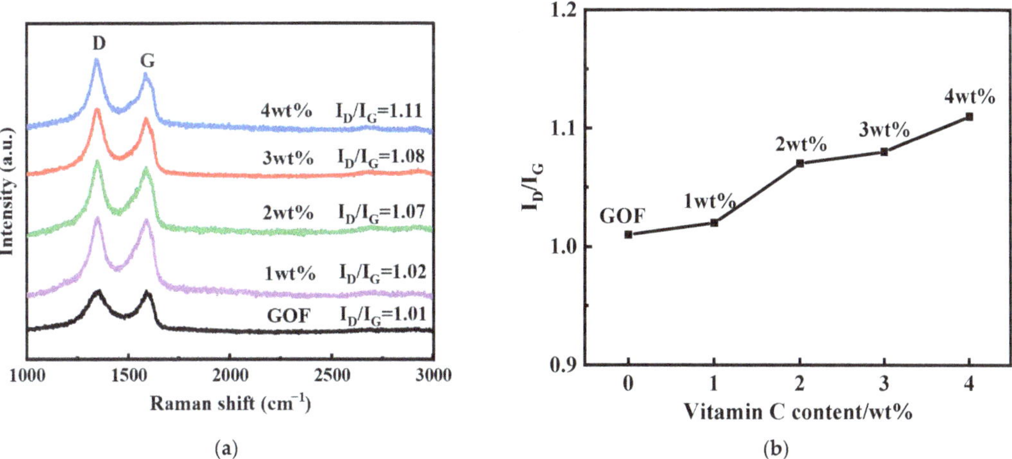

(**a**)   (**b**)

Figure 4. Raman spectra of GFs with different amounts of vitamin C doping. (**a**) Raman spectra of GOFs and GFs; (**b**) Comparison of Raman results between GOFs and GFs.

The XPS results of GOF and GF are presented in Figure 5. Analysis of Figure 5f reveals the presence of C1s (284 eV) and O1s (532 eV) based on the observed peaks. By performing deconvolution on the high-resolution C1s peak of GOF, it is observed that there are four peaks corresponding to C-C, C-O, C=O, and O-C=O, centered at 284.8 eV, 285.4 eV, 286.6 eV, and 288.7 eV, respectively. Upon annealing, GF exhibits a similar and significant decrease in these peaks, indicating the partial decomposition of oxygen functional groups and the partial recovery of π-conjugated structures after the addition of vitamin C. The profiles of GF with different levels of vitamin C were approximately the same. The differences in the C/O ratios were evident in the XPS full spectra, which were 2.58 for GOF, 3.80 for 1 wt%-GF, 4.33 for 2 wt%-GF, 4.56 for 3 wt%-GF, and 4.79 for 4 wt%-GF. After vitamin C reduction, the C/O ratios all increased, indicating the effectiveness of vitamin C on GOF deoxygenation. The largest C/O ratio was found for 4 wt%-GF, proving that the highest degree of fiber reduction was achieved with the addition of 4 wt% vitamin C.

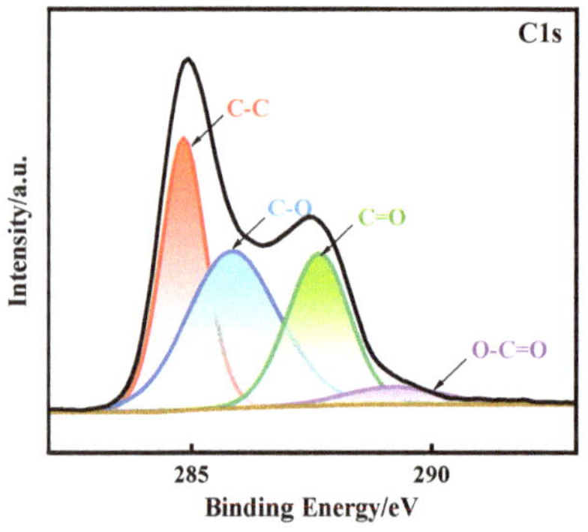

(**a**) XPS spectra of C1s of GOFs

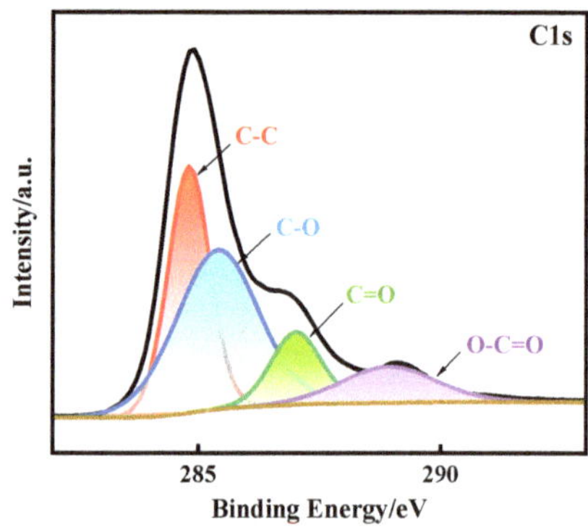

(**b**) XPS spectra of C1s of 1 wt%-GFs

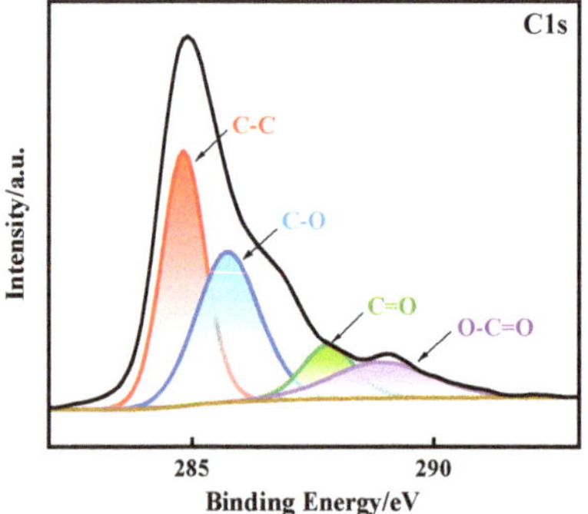

(**c**) XPS spectra of C1s of 2 wt%-GFs

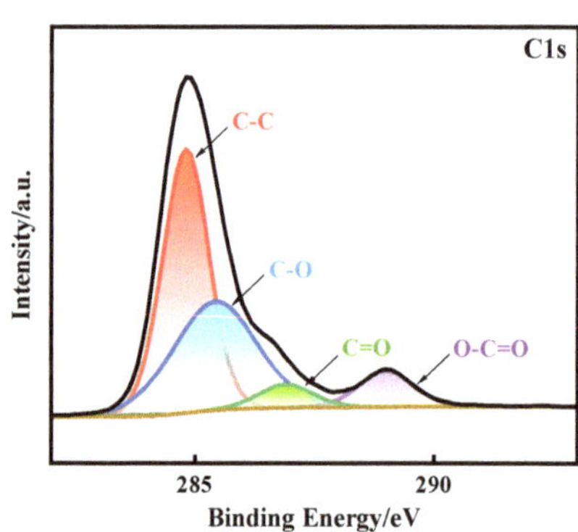

(**d**) XPS spectra of C1s of 3 wt%-GFs

Figure 5. *Cont.*

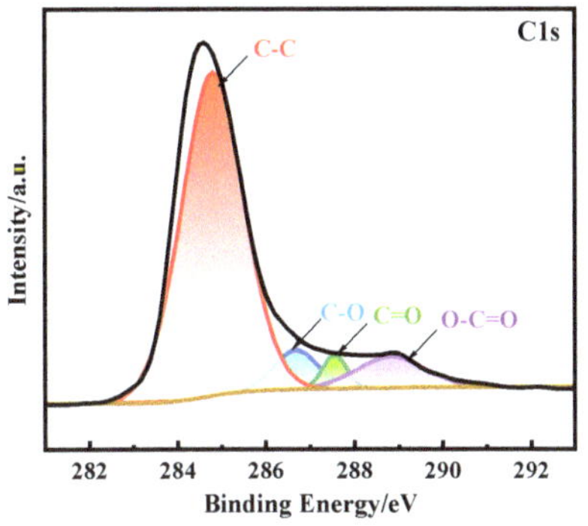

(**e**) XPS spectra of C1s of 4 wt%-GFs

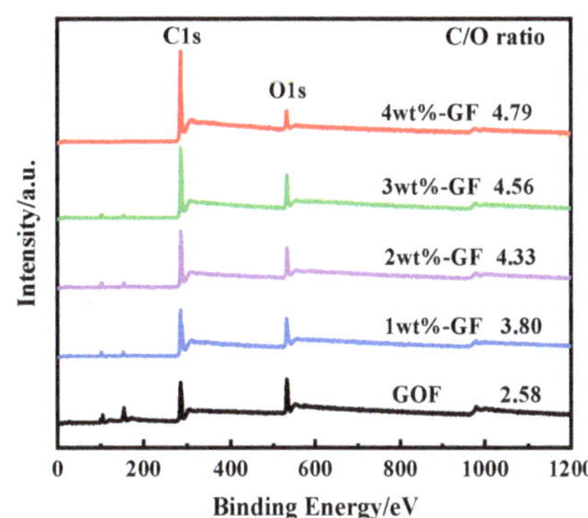

(**f**) Full XPS spectra of GOF and GF

Figure 5. XPS spectra of GFs with varying levels of vitamin C.

The effect of vitamin C on the degree of reduction of graphene fibers was analyzed in Figure 6a. The electrical conductivity of GOF is very poor (0.117 S/m), and that of GF rises rapidly after the addition of vitamin C. Through the increased vitamin C content, the electrical conductivities of GFs increase rapidly. The phenomenon occurs because the oxygen functional groups, including -OH, C-O-C, and -COOH, start to break down when vitamin C is reduced. This leads to the deoxygenation of sp^3 hybridized carbon atoms, allowing them to regain their sp^2 hybridized state and restore the conjugated C=C bond and π bond over time. Additionally, when the fiber is heat treated at 150 °C, the decomposition of oxygen-containing functional groups causes a gradual decrease in resistance and a rapid increase in conductivity. The 4 wt%-GF has the largest conductivity of 2448.8 S/m. Combined with the characterization analysis in the previous section, the trend of the conductivity is the same as that of the C/O ratio in the XPS analysis. The conductivity of graphene fiber is a direct criterion to judge its reduction effect, and in summary, it can be concluded that the treatment with vitamin C can significantly improve the conductivity of GFs and successfully reduce GOFs to obtain GFs.

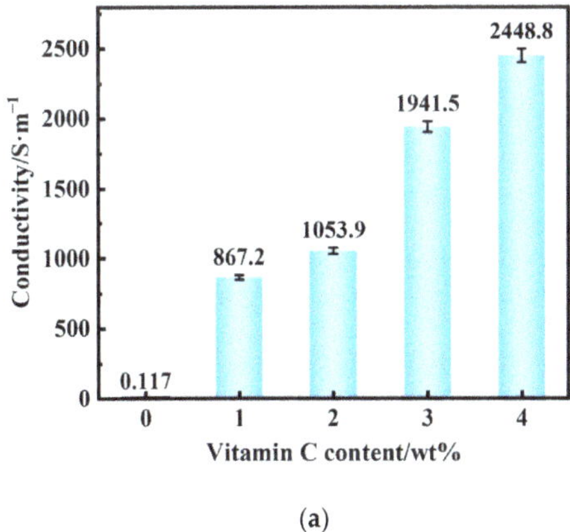

(**a**)

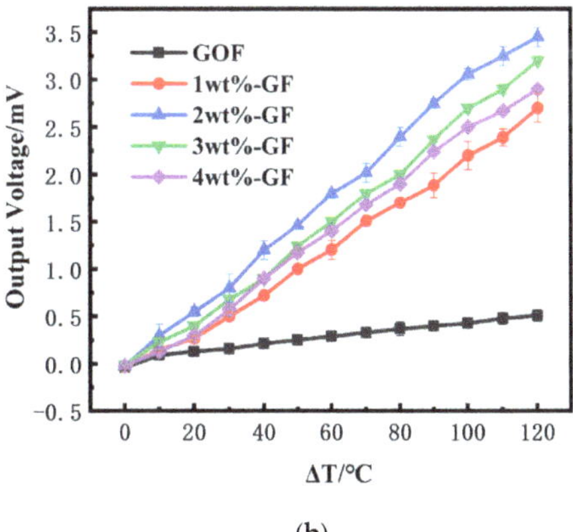

(**b**)

Figure 6. Electrical and thermoelectric properties of GFs. (**a**) Electrical conductivities of GFs at different vitamin C doping levels; (**b**) Thermoelectric output curve of GF/Pt thermocouple.

To investigate the thermoelectric characteristics of graphene fibers, a platinum wire was used as a reference electrode with GF at different vitamin C doping levels to form a thermocouple to test its thermoelectric properties. Since the Seebeck coefficient of a standard Pt wire is very small, about 1.67 $\mu V/°C$ [18], Pt can be used as one of the hot electrodes in the thermocouple, and GF at different annealing temperatures can be used as the other hot electrode. The output voltage characteristics of the thermocouple can be examined to qualitatively compare the influence of vitamin C doping on the thermoelectric performance of GF. An autonomous thermoelectric test system was used to test the performance of GOF/Pt and GF/Pt thermocouples, and the output voltages were surveyed by means of a temperature difference applied at both ends of the thermocouples under different steady-state temperature differences. The cold end of the thermocouple was placed in the air at about 25 °C. The output thermopotential of the thermocouple was recorded during the warming process as the temperature of the heating tape was regulated to increase the hot end temperature. The correlation between the output voltage of the thermocouple and the temperature difference between the two ends can be observed in Figure 6b. The thermopotential in the graph shows good linearity in the range of temperature differences of 0–120 °C. It can be concluded that the Seebeck coefficient of the material undergoes a significant enhancement after the reduction with vitamin C. The material has a very good thermal potential. The maximum output voltage of the thermocouple (3.45 mV) was observed at the introduction of 2 wt% of vitamin C. The output voltage of the thermocouple was also found to be higher than that of the original material. Due to the presence of oxygen impurities, the pristine graphene fiber exhibits p-type majority carriers and is a p-type thermoelectric material. As can be seen in Figure 6a, the electrical conductivity of GOF rises after the addition of vitamin C. This is because the oxygen functional groups in GF are reduced with the increasing addition of vitamin C. However, the output voltage of the thermocouple has not shown the same trend with the increase in vitamin C (shown in Figure 6b). According to Formula (1), the output voltage is determined by the Seebeck coefficient of materials. The Seebeck coefficient of GF is affected not only by conductivity but also by the Fermi level. As indicated in the Raman characterization (Figure 4a), more structural defects are introduced with the increasing content of vitamin C treatment. When the vitamin content increases from 0 to 2%, the Fermi level state of graphene is improved, resulting in a further increase in the Seebeck coefficient. When the vitamin content increases from 2 to 4%, excessive structural defects will affect the Fermi level state of graphene, leading to a decrease in the Seebeck coefficient. Thus, the output voltage will vary with changes in the Seebeck coefficient [19].

It is important to note that the measured thermal voltage output is influenced not only by the temperature difference between the hot and cold junctions but also by the Seebeck coefficient of the thermoelectric material. To further characterize the relation between thermal voltages and temperature differences, the results of fitting the thermoelectric output curves of the thermocouple temperature sensors with different vitamin C contents (fit R2 > 0.97) are shown in Table 1. The Seebeck coefficient of 2 wt%-GF/Pt was the largest (29.9 $\mu V/°C$). Vitamin C has the ability to eliminate the oxygen-containing functional groups present in GO. This process enhances the movement of charge carriers within the graphene layer, resulting in improved carrier mobility. Consequently, the Seebeck coefficient, which measures the conversion of temperature difference into electrical voltage, is enhanced.

It has been proven that the use of green and harmless reducing agents, such as tea extract, ethylene glycol, sugar, and vitamin C, is an environmentally friendly, mild, and effective method to obtain reduced graphene oxide (RGO) [20]. Vitamin C, also known as ascorbic acid [21], has the advantages of low cost, non-toxicity, and environmental friendliness.

The chemical reduction mechanism of GO is shown in Figure 7. The essence of GO chemical reduction is the reaction between the reducing agent and one or more functional groups on GO. Due to the different reactivity of each oxygen-containing functional group to GO, different reducing agents have different effects on removing functional groups. Moreover, vitamin C is rich in flavonoids, which react with the GO solution and undergo oxidation by the oxygen groups on the surface of the GO sheet. This reaction leads to the formation of benzoquinone products. Hydrogen ions can act as a catalyst during dehydration, where the epoxy group combines with hydrogen ions in solution to form a hydroxyl group, which in turn combines with the vacant half-bonds, resulting in the detachment of the water molecules and the carbon atoms of the rGO bonded to the sp^2 bond [22].

Table 1. Output fitting data of GF/Pt thermocouples with different vitamin C doping contents.

Sample	V~ΔT/μV	Seebeck Coefficient/μV·°C^{-1}	Correlation Coefficient (R^2)
GOF/Pt *	V = 28.13 + 4.16ΔT	4.16	0.977
1 wt%-GF/Pt	V = −136.92 + 23.04ΔT	23.04	0.995
2 wt%-GF/Pt	V = −23.30 + 29.90ΔT	29.90	0.997
3 wt%-GF/Pt	V = −104.95 + 27.27ΔT	27.27	0.996
4 wt%-GF/Pt	V = −123.52 + 25.57ΔT	25.57	0.996

* It represents the thermocouple with 0 wt% vitamin C added to GF as the anode and Pt as the cathode.

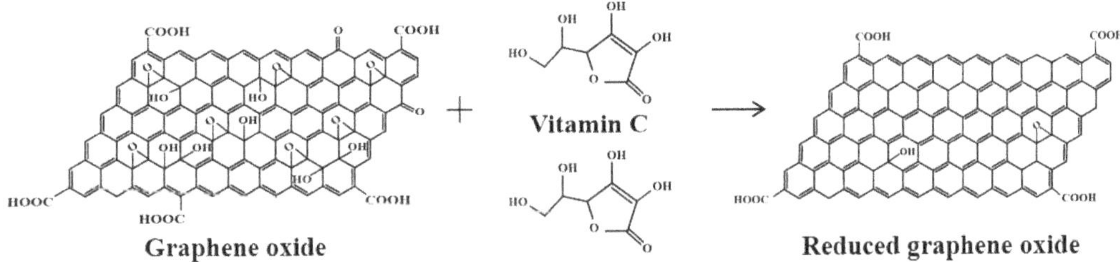

Figure 7. Mechanistic analysis diagram of chemical reduction of GO.

As shown in Figure 8a, the graphene fibers can be bent at any angle and can also be tied into a knot without the fibers breaking, and these results demonstrate the flexibility and torsion resistance of graphene fibers. For thermocouples, their cycle stability and repeatability are important performance parameters. Figure 8b,c showed plots of the output voltage of the thermocouple versus the temperature difference under three repetitions of the test. Fitting the test data over multiple warm-ups reveals that both the temperature difference and the output voltage are linear, and the data from the multiple tests overlap almost exactly, as does the cool-down process. Importantly, the slopes of the plots for multiple warm-ups and multiple cool-downs are almost identical, which indicates that the thermocouples have almost no drift in the test data during multiple warm-ups and cool-downs. The values of goodness of fit obtained by least squares fitting are all greater than 0.99, indicating a good linear relationship. This shows that the thermocouple has good cyclic stability and repeatability.

Figure 8d shows the bending and recovery processes of the fiber. The fibers were clamped on the slide table of the screw guide and were continuously bent and straightened with the rotation of the screw, leading to fixing one end of the fiber and bending back and forth the other end at a speed of 1 mm/s. The fiber was 6 cm long with a radius of curvature of 5 mm. As we can see from Figure 8e, the detecting sensitivities all remain basically unchanged after bending 1000 times in different flexible states (such as 30°, 60°, and 90° bending), suggesting that the thermocouple has the ability to be used for a long time in the bending process.

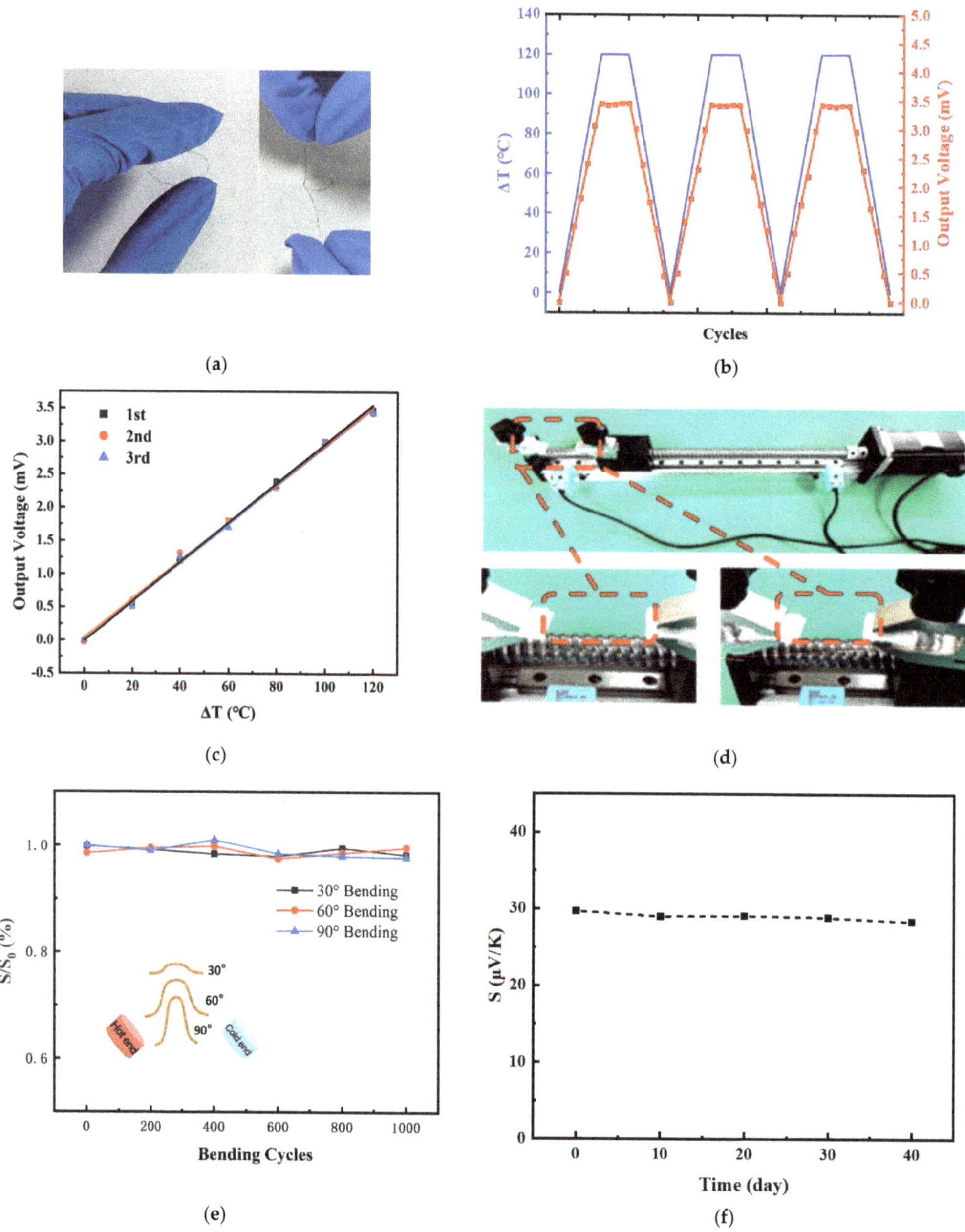

Figure 8. Characterization of GF electrode. (**a**) Bending and knotting diagram of graphene fibers; (**b**) 2 wt%-GF/Pt thermocouple repeatability test; (**c**) Thermoelectric output curves of the thermocouple for three warming processes; (**d**) Diagram of fiber bending and recovery process; (**e**) The sensitivity variation after bending 1000 times in different flexible states; (**f**) Plot of thermocouple sensitivity versus time.

Figure 8f shows the sensitivity of the fabricated thermocouples over time when left in the air. It can be seen that the sensitivity of the thermocouple encapsulated with PI film shows high stability (>95.66%) over 40 days.

Since the length of graphene composite fibers can be prepared as desired, temperature sensors with different pairs of p-n thermocouples can be easily fabricated. Figure 9 illustrates a plot of temperature difference versus output voltage for a temperature sensor with 8 pairs of thermocouples. The results show that the temperature sensor with 8 pairs of thermocouples has a sensitivity of about 200 µV/°C, which produces a substantial increase in sensitivity compared to the temperature sensor composed of 1 pair of thermocouples. Therefore, higher detection sensitivity can be obtained by increasing the number of thermocouple pairs comprising the temperature sensor.

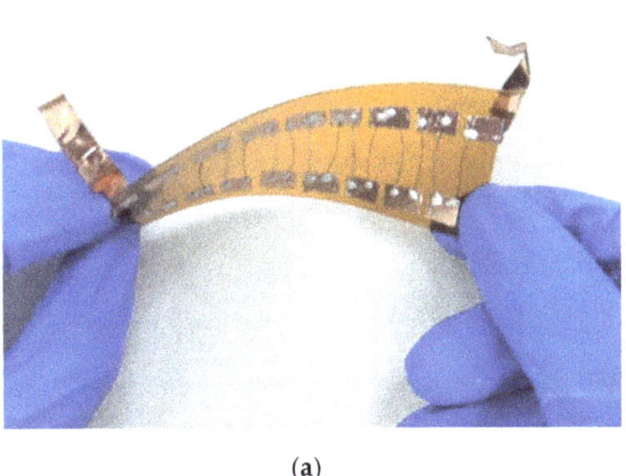

(a)

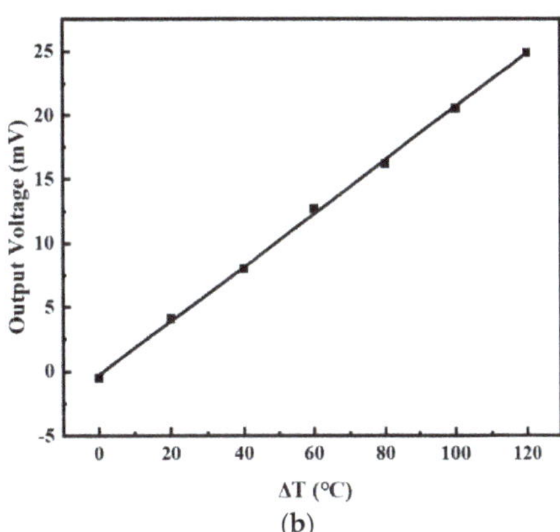

(b)

Figure 9. Performance test of multipair thermocouple temperature sensors. (**a**) A temperature sensor having 8 pairs of thermocouples. (**b**) The relationship between temperature difference and output voltage.

4. Conclusions

In this paper, GOFs were successfully fabricated by the microfluidic spinning technique. Vitamin C was employed as a reducing agent to reduce GOFs to GFs. The study focused on examining the impact of different concentrations of vitamin C (ranging from 1 wt% to 4 wt%) on the synthesis of GFs. To analyze the fibers, SEM maps and Raman spectroscopy were used to examine the surface morphology and internal structural defects. The electrical conductivity of 4 wt%-GFs is about 2448 S/m, which is 20,000 times higher compared to the electrical conductivity of pristine GOFs. The 2 wt%-GF and Pt were used to form a pair of thermocouple temperature sensors for testing, which showed good operating characteristics in the low-temperature measurement range with a sensitivity of up to 29.9 µV/°C. Additionally, it can be found that when there was a temperature difference of 120 °C between the cold and hot ends, the maximum thermoelectric output of the sensor reached 3.45 mV. The sensor also demonstrated resistance to bending interference and showed minimal change in sensitivity even after being bent 1000 times. The sensor encapsulated with PI film still stays considerably stable after being left in the air for 40 days. The sensitivity of the temperature sensor, consisting of 8 pairs of thermocouples, can be increased up to 200 µV/°C. With the benefits of small size, light weight, high sensitivity, and good scalability, graphene fiber-based flexible thermocouple temperature sensors are simple to make and environmentally friendly. They are anticipated to reflect human activities and health conditions via real-time monitoring of human body temperature and demonstrate a wide range of application potentials in the fields of e-skin and smart medicine.

Author Contributions: Conceptualization, C.W. and Y.Z.; methodology, F.H.; validation, C.W.; formal analysis, F.H.; investigation, Y.Z.; data curation, C.W.; writing—original draft preparation, Y.Z.; writing—review and editing, C.W. and F.H.; visualization, C.W. and Y.Z.; supervision, Z.J; project administration, Z.J.; funding acquisition, C.W., F.H., and Z.J. All authors have read and agreed to the published version of the manuscript.

Funding: This research was funded by the National Key Research and Development Program of China (No.2022YFB3205800) and the National Natural Science Foundation of China (Grant 52175434).

Data Availability Statement: Data sharing does not apply to this article.

Conflicts of Interest: The authors declare no conflict of interest.

References

1. Hammock, M.L.; Chortos, A.; Tee, B.C.-K.; Tok, J.B.-H.; Bao, Z. The evolution of electronic skin (e-skin): A brief history, design considerations, and recent progress. *Adv. Mater.* **2013**, *25*, 5997–6038. [CrossRef] [PubMed]
2. Garraud, A.; Combette, P.; Giani, A. Thermal stability of Pt/Cr and Pt/Cr2O3 thin-film layers on a SiNx/Si substrate for thermal sensor applications. *Thin Solid Film.* **2013**, *540*, 256–260. [CrossRef]
3. Huo, X.; Liu, H.; Liang, Y.; Fu, M.; Sun, W.; Chen, Q.; Xu, S. A nano-stripe based sensor for temperature measurement at the submicrometer and nano scales. *Small* **2014**, *10*, 3869–3875. [CrossRef] [PubMed]
4. Snyder, G.J.; Toberer, E.S. Complex thermoelectric materials. *Nat. Mater.* **2008**, *7*, 105–114. [CrossRef] [PubMed]
5. Bolotin, K.I.; Sikes, K.J.; Jiang, Z.; Klima, M.; Fudenberg, G.; Hone, J.; Kim, P.; Stormer, H.L. Ultrahigh electron mobility in suspended graphene. *Solid State Commun.* **2008**, *146*, 351–355. [CrossRef]
6. Novoselov, K.S.; Geim, A.K.; Morozov, S.V.; Jiang, D.; Zhang, Y.; Dubonos, S.V.; Grigorieva, I.V.; Firsov, A.A. Electric field effect in atomically thin carbon films. *Science* **2004**, *306*, 666. [CrossRef]
7. Weiss, N.O.; Zhou, H.; Liao, L.; Liu, Y.; Jiang, S.; Huang, Y.; Duan, X. Graphene: An emerging electronic material. *Adv. Mater.* **2012**, *24*, 5782–5825. [CrossRef]
8. Dragoman, D.; Dragoman, M. Giant thermoelectric effect in graphene. *Appl. Phys. Lett.* **2007**, *91*, 203116. [CrossRef]
9. Dong, Z.; Jiang, C.; Cheng, H.; Zhao, Y.; Shi, G.; Jiang, L.; Qu, L. Facile fabrication of light, flexible and multifunctional graphene fibers. *Adv. Mater.* **2012**, *24*, 1856–1861. [CrossRef]
10. Ugale, A.D.; Umarji, G.G.; Jung, S.H.; Deshpande, N.G.; Lee, W.; Cho, H.K.; Yoo, J.B. ZnO decorated flexible and strong graphene fibers for sensing NO2 and H2S at room temperature. *Sens. Actuators B Chem.* **2020**, *308*, 127690. [CrossRef]
11. Hummers, W.S.; Offerman, R.E. Preparation of Graphitic Oxide. *J. Am. Chem. Soc.* **1957**, *80*, 1339–1440. [CrossRef]
12. Martin, J.; Tritt, T.; Uher, C. High temperature Seebeck coefficient metrology. *J. Appl. Phys.* **2010**, *108*, 121101. [CrossRef]
13. Lerf, A.; He, H.; Forster, M.; Klinowski, J. Structure of graphite oxide revisited. *J. Phys. Chem. B* **1998**, *102*, 4477–4482. [CrossRef]
14. Bora, C.; Bharali, P.; Baglari, S.; Dolui, S.K.; Konwar, B.K. Strong and conductive reduced graphene oxide/polyester resin composite films with improved mechanical strength, thermal stability and its antibacterial activity. *Compos. Sci. Technol.* **2013**, *87*, 1–7. [CrossRef]
15. Zhang, J.; Yang, H.; Shen, G.; Cheng, P.; Zhang, J.; Guo, S. Reduction of graphene oxide via L-ascorbic acid. *Chem. Commun.* **2010**, *46*, 1112. [CrossRef]
16. Wang, J.; Chen, Z.M.; Chen, B.L. Adsorption of polycyclic aromatic hydrocarbons by graphene and graphene oxide nanosheets. *Environ. Sci. Technol.* **2014**, *48*, 4817–4825. [CrossRef] [PubMed]
17. Chang, K.; Li, X.; Liao, Q.; Hu, B.; Hu, J.; Sheng, G.; Linghu, W.; Huang, Y.; Asiri, A.M.; Alamry, K.A. Molecular insights into the role of fulvic acid in cobalt sorption onto graphene oxide and reduced graphene oxide. *Chem. Eng. J.* **2017**, *327*, 320–327. [CrossRef]
18. *ANSI/ASTM E230/E230M-2012*; Standard Specification and Temperature-Electromotive Force (emf) Tables for Standardized Thermocouples. ASTM International: West Conshohocken, PA, USA, 2017.
19. Yun, J.S.; Choi, S.; Im, S.H. Advances in carbon-based thermoelectric materials for high-performance, flexible thermoelectric devices. *Carbon Energy* **2021**, *3*, 667–708. [CrossRef]
20. Stobinski, L.; Lesiak, B.; Malolepszy, A.; Mazurkiewicz, M.; Mierzwa, B.; Zemek, J.; Jiricek, P.; Bieloshapka, I. Graphene oxide and reduced graphene oxide studied by the XRD, TEM and electron spectroscopy methods. *J. Electron Spectrosc. Relat. Phenom.* **2014**, *195*, 145–154. [CrossRef]
21. Mindivan, F.; Gkta, M. Effects of various vitamin C amounts on the green synthesis of reduced graphene oxide. *Mater. Test.* **2019**, *61*, 1007–1011. [CrossRef]
22. Wang, W.S.; Hou, D.D.; Liu, C.F.; Ding, S.L.; Xu, B.H. Mild reduction of graphene oxide by vitamin C and its characterization. *Carbon Technol.* **2018**, *37*, 6.

micromachines

Article

Structural Design of Dual-Type Thin-Film Thermopiles and Their Heat Flow Sensitivity Performance

Hao Chen [1,2], Tao Liu [3], Nanming Feng [1,2], Yeming Shi [1,2], Zigang Zhou [3] and Bo Dai [1,*]

[1] The State Key Laboratory of Environment-Friendly Energy Materials, Southwest University of Science and Technology, Mianyang 621010, China; chenhaoswustedu@163.com (H.C.); 18780550401@163.com (N.F.); seiyem@outlook.com (Y.S.)
[2] School of Materials and Chemistry, Southwest University of Science and Technology, Mianyang 621010, China
[3] School of Mathematics and Science, Southwest University of Science and Technology, Mianyang 621010, China; liut427619@163.com (T.L.); zhouzigang@swust.edu.cn (Z.Z.)
* Correspondence: daibo@swust.edu.cn; Tel./Fax: +86-0816-2480830

Abstract: Aiming at the shortcomings of the traditional engineering experience in designing thin-film heat flow meters, such as low precision and long iteration time, the finite element analysis model of thin-film heat flow meters is established based on finite element simulation methods, and a double-type thin-film heat flow sensor based on a copper/concentrate thermopile is made. The influence of the position of the thermal resistance layer, heat flux density and thickness of the thermal resistance layer on the temperature gradient of the hot and cold ends of the heat flow sensor were comprehensively analyzed by using a simulation method. When the applied heat flux density is $50\,\text{kW/m}^2$ and the thermal resistance layer is located above and below the thermopile, respectively, the temperature difference between the hot junction and the cold junction is basically the same, but comparing the two, the thermal resistance layer located above is more suitable for rapid measurements of heat flux at high temperatures. In addition, the temperature difference between the hot and cold contacts of the thin-film heat flux sensor increases linearly with the thickness of the thermal resistance layer. Finally, we experimentally tested the response–recovery characteristics of the sensors, with a noise of 2.1 µV and a maximum voltage output of 15 µV in a room temperature environment, respectively, with a response time of about 2 s and a recovery time of about 3 s. Therefore, the device we designed has the characteristic of double-sided use, which can greatly expand the scope of use and service life of the device and promote the development of a new type of heat flow meter, which will provide a new method for the measurement of heat flow density in the complex environment on the surface of the aero-engine.

Keywords: thin-film type; heat flow sensor; thermal resistance layer; thermopile; finite element simulation

Citation: Chen, H.; Liu, T.; Feng, N.; Shi, Y.; Zhou, Z.; Dai, B. Structural Design of Dual-Type Thin-Film Thermopiles and Their Heat Flow Sensitivity Performance. *Micromachines* **2023**, *14*, 1458. https://doi.org/10.3390/mi14071458

Academic Editors: Weidong Wang, Yong Ruan and Zai-Fa Zhou

Received: 25 June 2023
Revised: 12 July 2023
Accepted: 18 July 2023
Published: 20 July 2023

1. Introduction

With the rapid development of China's aerospace industry, the design and development of aero-engines, as the core components of aero-aircraft, must rely on advanced and effective testing technologies to verify their design performance indicators and reliability [1,2]. As the most important component of an aero-engine, the turbine blades have a narrow internal space and a harsh working environment, so it is necessary to accurately measure the surface temperature distribution and heat transfer process of high-temperature components such as turbines and combustion chambers to verify the cooling efficiency and performance of the thermal barrier coating, evaluate the combustion efficiency of the gas, monitor the engine running state, and troubleshoot [3–5]. Therefore, rapid and accurate measurement of the heat distribution and transfer characteristics on the surface of turbine blades of aero-engines is essential for the development of the aerospace industry.

A heat flux sensor is a kind of device or instrument that can detect and monitor the heat flux density in real time and convert it into an electrical signal for output [6–8]. It is widely used in the technical field of national defense, and the technology of measuring heat flux density has made great progress. In order to describe the harsh environmental changes of high-temperature airflow and high-pressure airflow on the turbine engine blades of the space shuttle more thoroughly, in addition to the temperature on the turbine engine blades as a parameter, heat flux density, as another important measurement parameter, has gradually begun gain interest the aviation technical field [9,10]. In order to solve the limitation of the heat flow sensor in high temperature and high pressure air flow environment, researchers have conducted extensive theoretical research and experimental verification. Traditional heat flow meters such as the Gordon heat flow meter [11], the Schmidt–Belter meter [12] and the standard stratification meter [13] are unable to meet the test requirements, such as being large, affecting the accuracy of heat flow measurements and being unsuitable in fields such as aerospace where a high measurement accuracy is required [14–17]. As a result, thin-film heat flow meters only began to appear in aerospace engine research in the 1990s, measuring a wide range of heat flows with a large output signal, making them one of the most widely used devices for measuring heat flow density today. Combined with MEMS (micro electro mechanical system) technology [18–20], the thin-film sensor is deposited on the surface of the measured object, which is integrated with the measured object, and has the characteristics of small volume, small heat capacity, low interference and no damage to the air flow on the surface of the component, which will also provide a new measurement method for the surface heat flow measurement of engine hot-end components [21,22].

In this paper, the finite element method is used to design and optimize the structure and position of the thermal resistance layer, and a MEMS thin-film heat flow sensor is prepared. The sensor is made of a copper/concentrate thin-film thermopile as the sensing structure, and SiO_2/Al_2O_3 as the bottom thermal resistance layer. The effect of the thermal resistance layer position, heat flow density and thickness of the thermal resistance layer on the temperature gradient of the cold and hot junction of the thermal flow sensor is investigated and the simulation results are compared with the experimental measurement results. The results of the simulations are compared with the experimental results to provide theoretical guidance and experimental validation for the further validation and development of a high-performance and stable heat flow sensor for monitoring heat flow density in the complex environment of aero-engine surfaces.

2. Working Principle

The working principle of a thin-film thermopile heat flow meter can be explained by heat conduction theory, as shown in Figure 1; that is, when there is heat flow through the sensor, the temperature changes only in the X direction, resulting in a temperature gradient. According to Fourier's law, the heat flow density through the heat flow meter can be obtained.

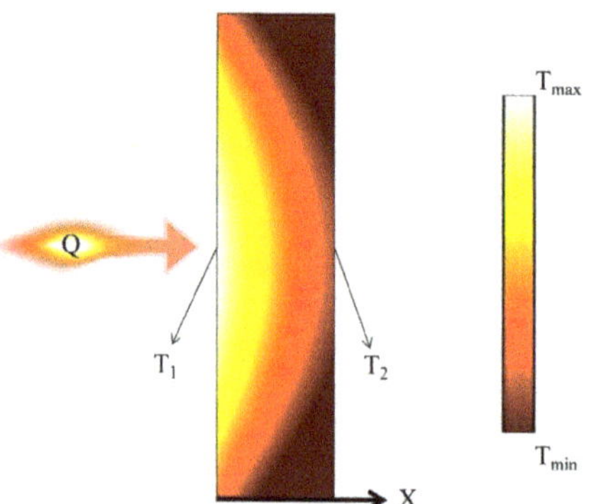

Figure 1. Schematic diagram of Fu Liye's law.

The heat flow density Q through the heat flow meter is expressed according to Equation (1) [22,23]:

$$Q = -\lambda \frac{\Delta T}{\Delta X} = -\lambda \frac{T_2 - T_1}{h} \tag{1}$$

where Q is the heat flow density, λ is the thermal conductivity of the thermal resistance layer, T_1 and T_2 are the temperatures of the upper and lower surfaces of the thermal resistance layer, respectively, and h is the thickness of the thermal resistance layer.

According to Equation (1), if the material and geometry of the device are determined, the magnitude of the heat flow density can also be obtained by measuring this temperature difference. And, according to the thermopile temperature measurement principle [24,25]:

$$E = S \cdot N \cdot \Delta T \tag{2}$$

where E is the magnitude of the voltage output, S is the Seebeck coefficient of the thermocouple and N is the number of pairs of thermocouples at the base of the device. Combining Equation (1) with Equation (2) yields the heat flow density versus thermopile output as Equation (3):

$$Q = \frac{\lambda}{S \cdot N \cdot h} E \tag{3}$$

With the above analysis, we can obtain the heat flux density by knowing the temperature difference of each thermocouple pair at the bottom or directly measuring the thermal potential.

3. Sensor Structure Design

As shown in Figure 2, according to the heat conduction theory and thermoelectric effect, the structural design of the dual-type thin-film heat flow sensor is schematically shown in Figure 2a, where the top structure is mainly composed of alternating Al_2O_3 and SiO_2 thermomechanical layers. Figure 2b shows the 3D structural diagram of the bottom view of the device, and below the thermal resistance layer is the thermopile layer (copper/concentrate thermopile), which is mainly used to convert the temperature difference on the lower surface of the resistance layer into a voltage for detection. It is worth noting that we have designed 200 pairs of thermocouples. This is because the voltage provided by a single pair of thermocouples is too small, so we have designed a reasonable structure to connect 200 pairs of thermocouples in series, which greatly increases the voltage output and reduces the influence of some accidental factors. The geometric parameters of the finite element model are shown in Table 1. In order to explain the working principle of the new heat flow meter we designed more clearly, we drew a side view of the device. When the heat flow is applied vertically to the sensor, due to the difference in thermal conductivity between the SiO_2 and Al_2O_3 thermal resistive layers, the Al_2O_3 thermal resistive layer with high thermal conductivity transfers the excess heat from the cold end to the Al_2O_3 ceramic substrate to form the cold end, while the SiO_2 thermal resistive layer with low thermal conductivity restricts the heat transfer from the hot end to the substrate to form the hot end, which creates a temperature gradient between the hot end and the cold end. The underlying temperature distribution of the bottom layer is also different, so we detect the temperature distribution of the bottom layer by adding a thermopile layer [26–28]. Therefore, it is possible to infer the heat flow in the top layer or the power of the laser by observing the output voltage.

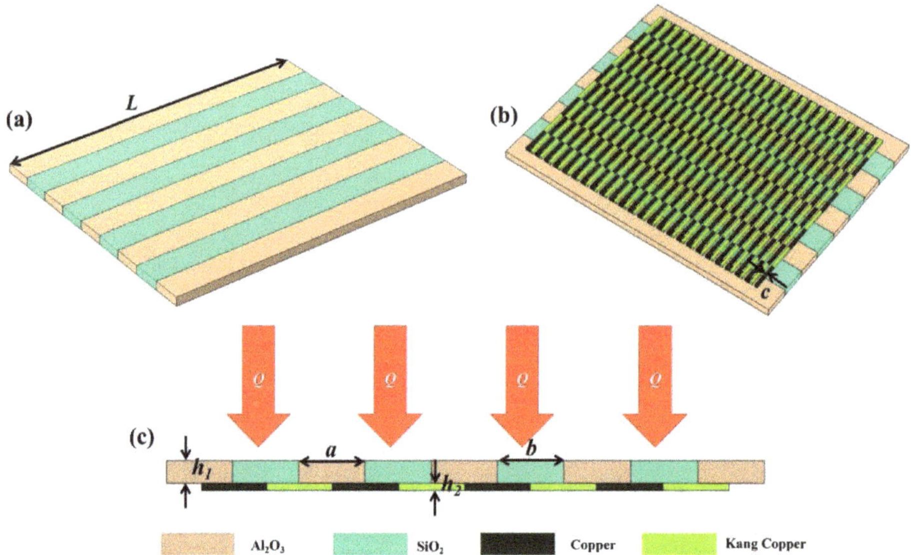

Figure 2. Structural diagram of the heat flow meter. (**a**) Front three-dimensional structure drawing. (**b**) Reverse three-dimensional structure. (**c**) Side view of the heat flow meter.

Table 1. Dual type thin film thermopile heat flow sensor structure parameter table.

Parameter Name	Description of Parameters	Parameter Values
Q	Heat flow density	$50 \, \text{KW/m}^2$
h_1	Thickness of thermal resistance layer	$5 \, \mu\text{m}$
h_2	Thermoelectric stack thickness	$1 \, \mu\text{m}$
L	Length of thermal resistance layer	$1 \, \mu\text{m}$
a	Width of Al_2O_3	$1000 \, \mu\text{m}$
b	Width of SiO_2	$1000 \, \mu\text{m}$
c	Thermocouple length	$100 \, \mu\text{m}$
N	Number of thermocouples	200

4. Finite Element Simulation Analysis

4.1. Effect of Different Thermal Resistance Layer Positions on Heat Flow Measurements

In order to ensure that the designed sensor can show good response characteristics, the finite element software of COMSOL version 5.4 was used for simulation analysis, a boundary heat source was added to simulate the heat flow, and the connection position between thermocouple and thermal resistance layers is meshed to obtain the theoretical response characteristics and sensitivity of the sensor and compared with experimental results. The structural parameters of the finite element model created by the simulation are the same as those of the prepared device. Figure 3a shows the temperature distribution curve of the thin-film heat flow sensor after reaching steady state when the thermal resistance layer is located above the thermopile design structure and the heat flow of $50 \, \text{KW/m}^2$ is vertically applied to the sensor. Figure 3b shows the temperature cloud on the sensor surface. It can be clearly observed that due to the difference in thermal conductivity between the SiO_2 and Al_2O_3 thermal resistance layers, a temperature difference is generated on the thermopile that leads to a voltage output, which indicates that the design of the structure of the thin-film heat flow sensor is reasonable. It can also be seen that the temperature below Al_2O_3 layer is higher, with an average temperature at the hot end of 298.704 K, and the temperature below SiO$_2$ layer is lower, with an average temperature at the cold end of 298.663 K and a temperature difference of 0.0409 K. It is worth mentioning that the average temperature of the hot end when the thermal resistor layer is located above the thermopile is lower than that when it is located below, which is more suitable for rapid measurement of heat flow at low temperatures.

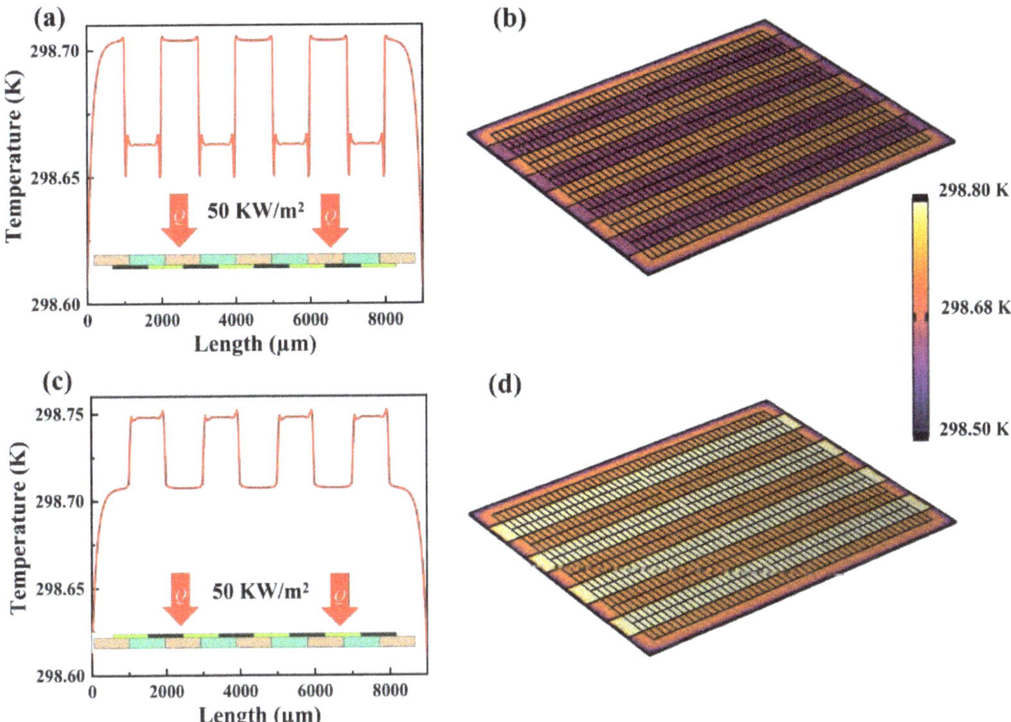

Figure 3. Temperature distribution profile when applying a heat flow density of 50 KW/m². (**a**) Temperature distribution profile on the thermopile layer for frontal incidence of heat flow. (**b**) Temperature cloud over the thermopile layer at frontal incidence. (**c**) Temperature profile on the thermopile layer at frontal incidence. (**d**) Temperature cloud over the thermopile layer at frontal incidence.

Figure 3c shows the temperature distribution curve of the thin-film heat flow sensor when a heat flow of 50 KW/m² is applied vertically to the sensor after reaching steady state when the heat resistance layer is located below the thermopile design structure, and Figure 3d shows the temperature cloud on the surface of the sensor. It can be seen that the temperature at the hot end of the thermocouple is roughly 298.748 K, while the average temperature at the cold end is 298.707 K, with a temperature difference of 0.041 K.

According to the Seebeck effect, the output voltage when the thermal resistance layer is located below the thermopile design structure is 13.1 μV. This is because the carriers within the conductor move from the hot end to the cold end under the temperature gradient, and accumulate at the cold end, thus generating a potential difference inside the material, and under this potential difference, a reverse charge flow is generated. When the thermal charge flow and the internal electric field are in dynamic balance, a stable temperature difference potential is formed at both ends of the material. The smaller the thickness of the thermal resistance layer, the shorter the heat transfer time and the faster the response time.

4.2. Effect of Different Thermal Resistance Layer Thicknesses on Heat Flow Measurements

The thickness of the thermal resistance layer of a thermoelectric stack-type heat flow sensor has a huge impact on its performance. The greater the thickness, the longer the heat transfer time, and the greater the change of the temperature distribution on the lower surface. In order to study the influence of thermal resistance layers with different thicknesses (3, 4, 5, 6, 7 and 8 µm) on the performance, simulations were carried out. With a steady-state heat flow of 50 KW/m^2 applied, Figure 4a shows the temperature distribution profile of the lower surface of the thermal resistance layer at different thicknesses. The temperature at the hot end remains almost constant at around 298.704 K, but the temperature at the cold end increases with increasing thickness of the thermal resistance layer. Figure 4b shows the pattern of temperature difference between the hot and cold junctions with the thickness of the thermal resistance layer. When the thickness of the thermal resistance layer is 3 µm, the temperature difference between the hot and cold junctions is 0.008 K, but when the thickness is increased to 8 µm, the temperature difference increases to 0.16 K. As the thickness of the thermal resistance layer increases, the temperature at the hot end of the thin-film heat flow sensor remains constant, the temperature at the cold end increases linearly and the temperature difference between the hot and cold junctions also increases linearly, indicating that by increasing the thickness of the thermal resistance layer, the sensitivity of the device can be effectively improved. But, at the same time, the response time requirement must also be considered. Therefore, when choosing the thickness of the thermal resistance layer, we need to consider the balance between sensitivity and response time, because as the thickness of the thermal resistance layer increases, the heat transfer time and response time will be prolonged [29,30]. In a word, the heat flow sensitivity of the sensor can be effectively improved by reducing the thickness of the thermal resistance layer. However, we should also note that in order to meet the specific requirements and optimize the performance of the sensor, it is necessary to consider the corresponding reduction in heat transfer time, equipment heat transfer protection and response time requirements in combination with these factors.

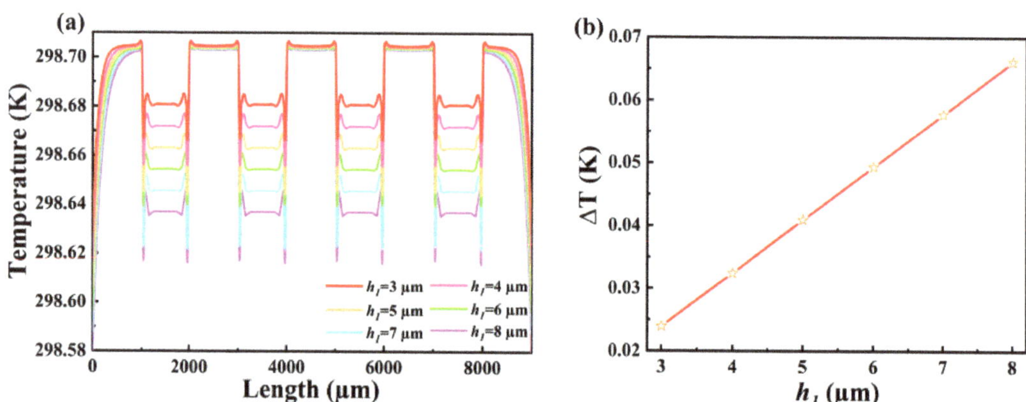

Figure 4. Variation of (**a**) temperature distribution curve and (**b**) temperature difference of individual thermopiles for thermopile layers with different thicknesses of thermal resistance layer.

4.3. Effect of Different Heat Flow Densities on Heat Flow Measurements

The devices we designed are mainly used to detect the magnitude of heat flow density or laser power density, so we investigated the change in the temperature of the bottom layer by varying the Q value applied to the top layer of the thermal resistive layer [31]. Figure 5 shows the temperature distribution on the sensor surface when the continuous heat flow of 10, 50, 100, 150 and 200 KW/m^2 acts on the sensor. As can be seen in Figure 5a, the overall temperature distribution of the device increases with increasing Q as the Q value varies from 10 to 200 KW/m^2. When Q = 100 KW/m^2, the hot end temperature is 304.258 K, and the cold end temperature is 304.176 K, so the stable temperature difference between the hot and cold end is 0.082 K. When Q = 100 KW/m^2, the temperature difference between the hot and cold end is ΔT = 0.163 K. Therefore, as shown in Figure 5b, the device Q value versus the corresponding ΔT, in which the corresponding ΔT increases as the i Q value increases, shows a good linear relation. Therefore, the thin-film heat flux sensor structure has a good output in the simulation test and can realize the accurate measurement of heat flux density. According to our simulation results, it is shown that our thin-film thermal flow sensor has good output precision and accuracy in measuring the heat flux. In a word, our research shows that by detecting the temperature change of the bottom layer, the accurate measurement of the heat flux can be realized. Our thin-film heat flux sensor structure shows good performance in simulation tests and can provide accurate heat flux measurements.

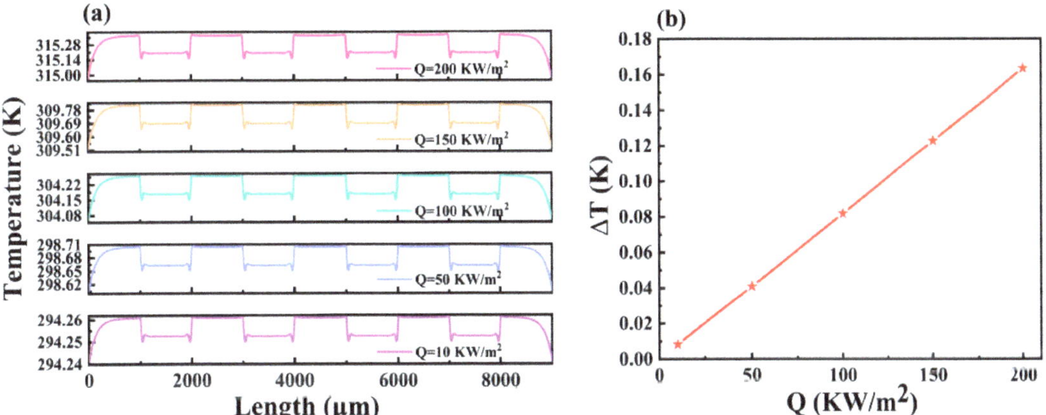

Figure 5. Temperature distribution profiles on (**a**) the thermopile layers and (**b**) the temperature difference (ΔT) variation on individual thermopiles when different heat flow densities are applied.

5. Fabrication and Performance Testing of Heat Flow Sensors

5.1. Heat Flow Sensor Production

Figure 6 shows the flow chart of the physical preparation of the designed heat flow meter. The main four coating processes were used to complete the fabrication [30,32], and the specific operation steps are shown in Figure 7: ① prepare the 4-inch silicon wafer for ultrasonic cleaning with deionized water, acetone and anhydrous ethanol for 15 min each; ② the first coating material is Al_2O_3, and then the final Al_2O_3 thermal resistance layer is formed through glue dumping, exposure, development and etching; ③ the second coating material is SiO_2, and the final SiO_2 layer is formed through glue dumping, exposure, development, ion etching and acetone debinding; ④ the above two coating processes are repeated to form the thermopile layer; and ⑤ finally, the metal electrodes are etched and soldered. The prepared sensor is shown in Figure 8, where the structural dimensions of the sensor are shown in Figure 8a,b, which show an optical microscope view of several interconnected thermopile strips. The thermopile layer (copper and con-copper layer) mainly uses the glow discharge principle to make argon gas ionization under high pressure after the formation of positive ions and bombardment of the target surface, so that the target particles sputtered out to reach the surface of the substrate to form a thermopile film of copper and con copper [33,34].

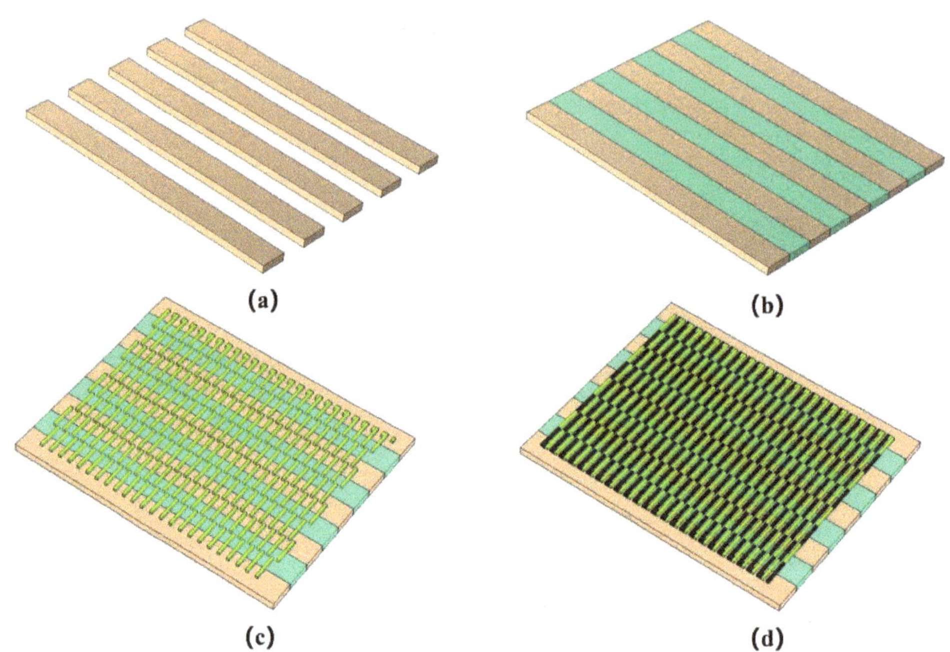

Figure 6. Schematic diagram of the preparation process. (**a**) Deposition of the Al$_2$O$_3$ thermal resistance layer. (**b**) Deposition of SiO$_2$ thermal resistance layer. (**c**) Deposition of the copper-containing layer of the thermopile. (**d**) Deposition of the copper layer of the thermopile.

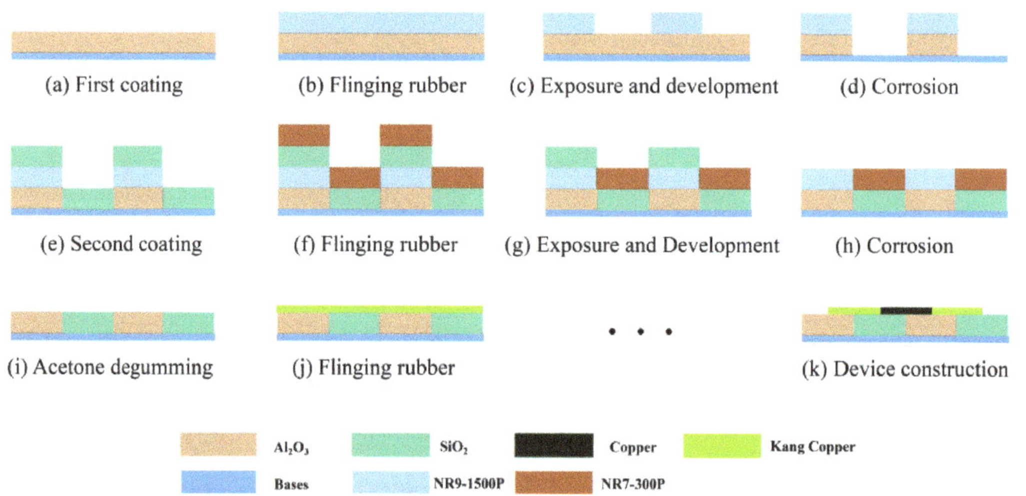

Figure 7. Process design flow diagram for the preparation of thin-film thermopiles.

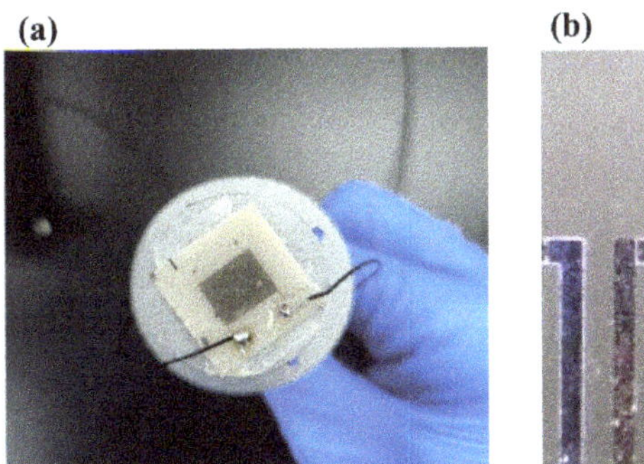

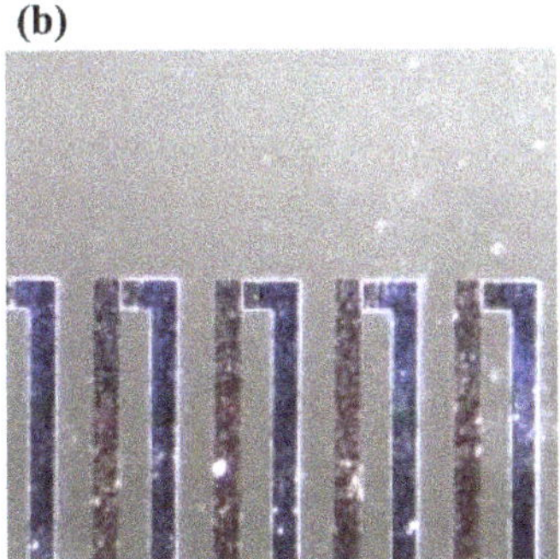

Figure 8. (**a**) Physical drawing of the structural dimensions of the sensor. (**b**) An optical microscope view of the thermopile.

5.2. Performance Testing of Heat Flow Sensors

As shown in Figure 9a, this study builds a high-temperature steady state and transient heat flow calibration system and conducts relevant experiments through the calibration platform and an infrared lamp to measure the sensitivity and reliability of the sensors, and then effectively evaluate the performance of the new thin-film-type thermo-fluid mete. Here we have chosen quartz infrared lamps as a highly efficient heating source, which has a quartz bubble shell and tungsten filament composition to avoid the blackening of the tube wall and the consequent decrease in light flux. Advantages of quartz infrared lamps include: high efficiency, fast heat transfer, and fast responses. Figure 9b shows the dynamic response curves of the prepared sensor samples in terms of the heat flow density of the IR lamp radiation. It can be seen that the sample exhibits excellent response–recovery characteristics with a noise of 2.1 μV and a maximum voltage output of 15 μV at room temperature, respectively, with a response time of about 2 s and a recovery time of about 3 s. This is in good agreement with the above simulation results. Wherein, the response time is the time from the exposure of the thermofluid meter to a heat flow of a given power to the output stabilizing the output voltage under test conditions; typically, the time to reach 90% of the maximum value is read as the response time. Whereas, the recovery time is the time from when the thermofluid meter is no longer in contact with the heat flow to when the output returns to a steady state under test conditions; typically, the time to recover to 10% of the maximum value is read as the recovery time [35,36]. In addition, the magnitude of the output voltage shows good agreement with the simulation calculations.

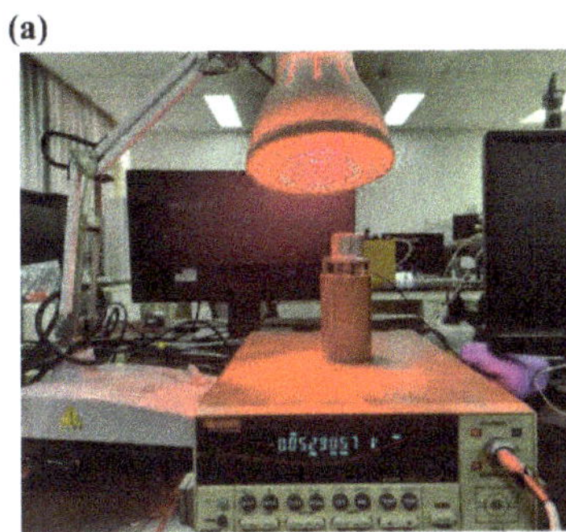

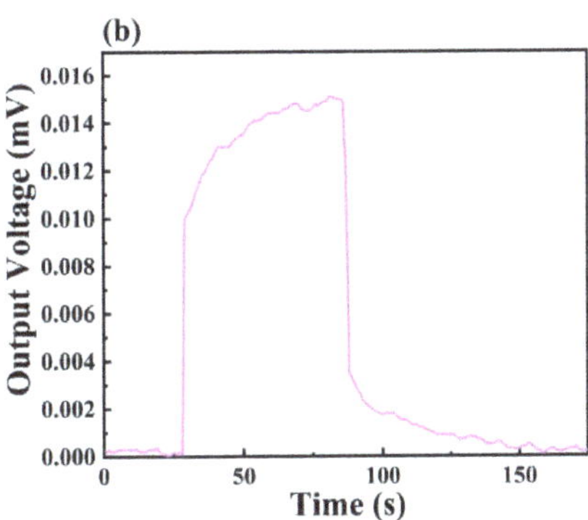

Figure 9. (**a**) Physical diagram of the experimental test setup. (**b**) Output voltage–time dynamic variation curve of the sample under IR lamp radiation.

Figure 10 shows the static response curves of the device under different heat flow densities, and it can be seen that the heat flow densities are in the range of 0–160 kW/m^2, and the output voltages are 0.13 mV, 0.27 mV, 0.27 mV, 0.39 mV, 0.93 mV, 2.25 mV, and 3.58 mV, respectively, when the heat flow densities are, in order, 20 kW/m^2, 25 kW/m^2, 30 kW/m^2, 50 kW/m^2, 100 kW/m^2, and 150 kW/m^2, and the corresponding output voltages are 0.39 mV, 0.93 mV, 2.25 mV and 3.58 mV, indicating that the sensor output thermoelectric potential shows a good linear relationship with heat flow. The test results were linearly fitted to obtain the binary sublinear equation, y = 0.02648x − 0.39679, the slope of which is the sensor sensitivity, which is 0.02648 mV/(kW/m^2).

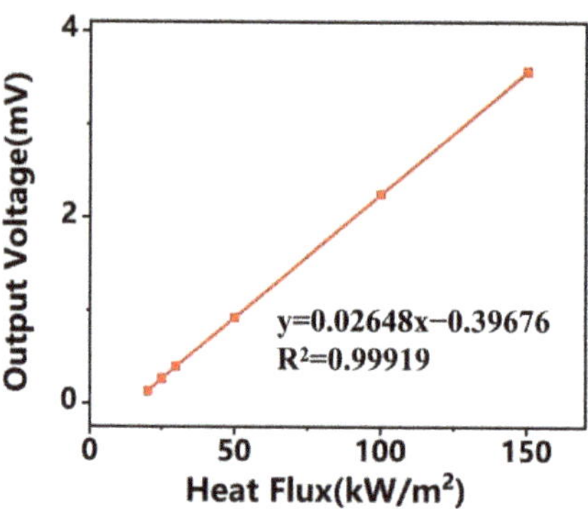

Figure 10. Steady-state calibration measurements of a dual-type thin-film heat flow meter.

6. Conclusions

In this paper, a double-structure thin-film heat flow sensor was designed, and a finite element model is established. The output voltage is adjusted by the relationship between the position of the thermal resistance layer and the thermopile, and the temperature difference between the hot end and cold ends of the heat flow sensor remains basically constant under the condition of the applied heat flow density of 50 kW/m^2. As the thickness of the thermoresistive layer increases, the cold-hot junction temperature difference of the thin-film thermofluid sensor increases linearly, from 0.008 K to 0.163 K, as the thickness of the thermoresistive layer increases from 3 μm to 8 μm. In addition, the results of the finite element simulation are compared with the experimental results. The results show that the sample has good response–recovery characteristics, the noise is 2.1 μV at room temperature, the maximum voltage output is 15 μV, the response time is about 2 s, and the recovery time is about 3 s, which is in good agreement with the above simulation results. The preparation process has good repeatability, which provides good technical support for the monitoring of heat flux density, standardization and mass production of aero-engine surface in a complex environment.

Author Contributions: H.C.: conceptualization, formal analysis, investigation, data curation, writing—original draft, and writing—review and editing. T.L.: conceptualization, formal analysis, investigation, data curation, and funding acquisition. N.F.: conceptualization, formal analysis, investigation, data curation, writing—original draft, and writing—review and editing. Y.S.: conceptualization, formal analysis, investigation, and data curation. Z.Z.: conceptualization, investigation, data curation, and writing—original draft. B.D.: conceptualization, formal analysis, investigation, data curation, and writing—original draft. All authors have read and agreed to the published version of the manuscript.

Funding: The authors are grateful to the support by the National Natural Science Foundation of China (11875228); the Project of State Key Laboratory of Environment-Friendly Energy Materials, Southwest University of Science and Technology (No. 20fksy23, 21fksy27, 22fksy26); the Southwest University of Science and Technology Provincial Student Innovation and Entrepreneurship Training Program (S202210619105); the Southwest University of Science and Technology Innovation Fund Precision Funding Special Project (JZ22-087); and the Southwest University of Science and Technology School of Mathematics and Science Innovation Fund Project (LX20210045, LX20210038).

Institutional Review Board Statement: Not applicable.

Informed Consent Statement: Not applicable.

Data Availability Statement: Data underlying the results presented in this paper are not publicly available at this time but may be obtained from the authors upon reasonable request.

References

1. Weng, H.; Duan, F.; Ji, Z.; Chen, X.; Yang, Z.; Zhang, Y.; Zou, B. Electrical insulation improvements of ceramic coating for high temperature sensors embedded on aeroengine turbine blade. *Ceram. Int.* **2020**, *46*, 3600–3605. [CrossRef]
2. Zhang, L.; Li, J.; Kou, Y. Research status of aero-engine blade fly-off. *J. Phys. Conf. Ser.* **2021**, *1744*, 22124–22125. [CrossRef]
3. Ji, Z.; Duan, F.; Xie, Z. Transient Measurement of Temperature Distribution Using Thin Film Thermocouple Array on Turbine Blade Surface. *IEEE Sens. J.* **2020**, *21*, 207–212. [CrossRef]
4. Ji, J.; Yan, B.; Wang, B.; Ding, M. Error of Thermocouple in Measuring Surface Temperature of Blade with Cooling Film. *IEEE Trans. Instrum. Meas.* **2022**, *71*, 1–12. [CrossRef]
5. Mersinligil, M.; Desset, J.; Brouckaert, J. High-temperature high-frequency turbine exit flow field measurements in a military engine with a cooled unsteady total pressure probe. *Proc. Inst. Mech. Eng. A J. Power Energy* **2011**, *225*, 954–963. [CrossRef]
6. Ribeiro, L.; Saotome, O.; d'Amore, R.; de Oliveira Hansen, R. High-Speed and High-Temperature Calorimetric Solid-State Thermal Mass Flow Sensor for Aerospace Application: A Sensitivity Analysis. *Sensors* **2022**, *22*, 3484. [CrossRef]
7. Wu, C.; Kang, D.; Chen, P.; Tai, Y. MEMS thermal flow sensors. *Sens. Actuators A Phys.* **2016**, *241*, 135–144. [CrossRef]
8. Fu, X.; Lin, Q.; Peng, Y.; Liu, J.; Yang, X.; Zhu, B.; Ouyang, J.; Zhang, Y.; Xu, L.; Chen, S. High-Temperature Heat Flux Sensor Based on Tungsten–Rhenium Thin-Film Thermocouple. *IEEE Sens. J.* **2020**, *20*, 10444–10452. [CrossRef]
9. Luo, X.; Wang, H. A High Temporal-Spatial Resolution Temperature Sensor for Simultaneous Measurement of Anisotropic Heat Flow. *Materials* **2022**, *15*, 5385. [CrossRef]
10. Balakrishnan, V.; Phan, H.-P.; Dinh, T.; Dao, D.V.; Nguyen, N.-T. Thermal Flow Sensors for Harsh Environments. *Sensors* **2017**, *17*, 2061. [CrossRef]
11. Fu, T.; Zong, A.; Zhang, Y.; Wang, H. A method to measure heat flux in convection using Gardon gauge. *Appl. Therm. Eng.* **2016**, *108*, 1357–1361. [CrossRef]
12. *ASTM E511-07*; Standard Test Method for Measuring Heat Flux Using a Copper-Constantan Circular Foil, Heat-Flux Transducer. ASTM International: West Conshohocken, PA, USA, 2007.
13. Matson, M.L.; Reinarts, T.R. Development and calibration of a thermal model of a Schmidt-Boelter gauge to quantify environmental and installation effects on heat flux measurements for launch vehicles. *Am. Inst. Phys.* **2002**, *608*, 119–126.
14. Sheng, C.; Hua, P.; Cheng, X. Design and calibration of a novel transient radiative heat flux meter for a spacecraft thermal test. *Rev. Sci. Instrum.* **2016**, *87*, 064902. [CrossRef]
15. Song, S.; Wang, Y.; Yu, L. Highly sensitive heat flux sensor based on the transverse thermoelectric effect of YBa2Cu3O7$-\delta$ thin film featured. *Appl. Phys. Lett.* **2020**, *117*, 123902. [CrossRef]
16. Zhang, T.; Tan, Q.; Lyu, W.; Xiong, J. Design and Fabrication of a Thick Film Heat Flux Sensor for Ultra-High Temperature Environment. *IEEE Access* **2019**, *7*, 180771–180778. [CrossRef]
17. Zhang, C.; Huang, J.; Li, J.; Yang, S.; Ding, G.; Dong, W. Design, fabrication and characterization of high temperature thin film heat flux sensors. *Microelectron. Eng.* **2019**, *217*, 111128. [CrossRef]
18. Sin, L.; Pan, M.; Tsai, T.; Chou, C. Multifunction thermopile sensors fabricated with a MEMS-compatible process. *IEEE Trans. Semicond. Manuf.* **2013**, *26*, 242–247. [CrossRef]
19. Morisaki, M.; Minami, S.; Miyazaki, K.; Yabuki, T. Direct local heat flux measurement during water flow boiling in a rectangular minichannel using a MEMS heat flux sensor—ScienceDirect. *Exp. Therm. Fluid Sci.* **2020**, *121*, 110285. [CrossRef]
20. Dejima, K.; Nakabeppu, O. Local instantaneous heat flux measurements in an internal combustion engine using a MEMS sensor. *Appl. Therm. Eng.* **2022**, *201*, 117747. [CrossRef]
21. Siroka, S.; Berdanier, R.A.; Thole, K.A.; Chana, K.; Haldeman, C.W.; Anthony, R.J. Comparison of thin film heat flux gauge technologies emphasizing continuous-duration operation. *J. Turbomach.* **2020**, *142*, 091001. [CrossRef]
22. Cui, Y.; Liu, H.; Wang, H.; Guo, S.; E, M.; Ding, W.; Yin, J. Design and Fabrication of a Thermopile-Based Thin Film Heat Flux Sensor, Using a Lead—Substrate Integration Method. *Coatings* **2022**, *12*, 1670. [CrossRef]

23. Singh, S.; Yadav, M.; Khandekar, S. Measurement issues associated with surface mounting of thermopile heat flux sensors. *Appl. Therm. Eng.* **2017**, *114*, 1105–1113. [CrossRef]
24. Prangemeier, T.; Nejati, I.; Müller, A.; Endres, P.; Fratzl, M.; Dietzel, M. Optimized thermoelectric sensitivity measurement for differential thermometry with thermopiles. *Exp. Therm. Fluid Sci.* **2015**, *65*, 82–89. [CrossRef]
25. E, M.; Cui, Y.; Guo, S.; Liu, H.; Ding, W.; Yin, J. Novel lead-connection technology for thin-film temperature sensors with arbitrary electrode lengths. *Meas. Sci. Technol.* **2023**, *34*, 065113. [CrossRef]
26. Wang, M.; Chen, J. Numerical analysis of influence factor on convective heat transfer coefficient about nozzle internal flow. *J. Rocket. Propuls.* **2011**, *37*, 32–37.
27. Abouelregal, A.; Tiwari, R.; Nofal, T. Modeling heat conduction in an infinite media using the thermoelastic MGT equations and the magneto-Seebeck effect under the influence of a constant stationary source. *Arch. Appl. Mech.* **2023**, *93*, 2113–2128. [CrossRef]
28. Hadi, A.-S.; Hill, B.E.; Issahaq, M.N. Performance Characteristics of Custom Thermocouples for Specialized Applications. *Crystals* **2021**, *11*, 377. [CrossRef]
29. Zheng, Z.; Luo, Y.; Yang, H.; Yi, Z.; Zhang, J.; Song, Q.; Yang, W.; Liu, C.; Wu, X.; Wu, P. Thermal tuning of terahertz metamaterial properties based on phase change material vanadium dioxide. *Phys. Chem. Chem. Phys.* **2022**, *24*, 8846–8853. [CrossRef]
30. Zheng, Y.; Yi, Z.; Liu, L.; Wu, X.W.; Liu, H.; Li, G.F.; Zeng, L.C.; Li, H.L.; Wu, P.H. Numerical simulation of efficient solar absorbers and thermal emitters based on multilayer nanodisk arrays. *Appl. Therm. Eng.* **2023**, *230*, 120841. [CrossRef]
31. Chen, H.; Li, W.; Zhu, S.; Hou, A.; Liu, T.; Xu, J.; Zhang, X.; Yi, Z.; Yi, Y.; Dai, B. Study on the Thermal Distribution Characteristics of a Molten Quartz Ceramic Surface under Quartz Lamp Radiation. *Micromachines* **2023**, *14*, 1231. [CrossRef]
32. Sharma, E.; Rathi, R.; Misharwal, J.; Sinhmar, B.; Kumari, S.; Dalal, J.; Kumar, A. Evolution in Lithography Techniques: Microlithography to Nanolithography. *Nanomaterials* **2022**, *12*, 2754. [CrossRef] [PubMed]
33. Meisak, D.; Plyushch, A.; Macutkevič, J.; Grigalaitis, R.; Sokal, A.; Lapko, K.N.; Selskis, A.; Kuzhir, P.P.; Banys, J. Effect of temperature on shielding efficiency of phosphate-bonded $CoFe_2O_4$–$xBaTiO_3$ multiferroic composite ceramics in microwaves. *J. Mater. Res. Technol.* **2023**, *24*, 1939–1948. [CrossRef]
34. Duan, F.; Chen, K.; Yu, Y. High-speed and low-power thermally tunable devices with suspended silicon waveguide. *Opt. Quantum Electron.* **2020**, *52*, 5. [CrossRef]
35. Mineo, G.; Moulaee, K.; Neri, G.; Mirabella, S.; Bruno, E. Mechanism of Fast NO Response in a WO3-Nanorod-Based Gas Sensor. *Chemosensors* **2022**, *10*, 492. [CrossRef]
36. Sheikhnejad, Y.; Bastos, R.; Vujicic, Z.; Shahpari, A.; Teixeira, A. Laser Thermal Tuning by Transient Analytical Analysis of Peltier Device. *IEEE Photonics J.* **2017**, *93*, 1–13. [CrossRef]

 *micromachines*

Article

Effect of Annealing Time on the Cyclic Characteristics of Ceramic Oxide Thin Film Thermocouples

Yuning Han [1,†], Yong Ruan [2,*,†], Meixia Xue [3], Yu Wu [2,4], Meng Shi [5], Zhiqiang Song [5], Yuankai Zhou [5] and Jiao Teng [3]

[1] Department of Electronic Information, Beijing Information Science and Technology University, Beijing 100192, China
[2] Department of Precision Instruments, Tsinghua University, Beijing 100084, China
[3] Department of Materials Physics and Chemistry, University of Science and Technology Beijing, Beijing 100083, China
[4] Qiyuan Laboratory, Beijing 100094, China
[5] MEMS Institute of Zibo National High-Tech Industrial Development Zone, Zibo 255000, China
* Correspondence: ruanyong@mail.tsinghua.edu.cn
† These authors contributed equally to this work.

Abstract: Oxide thin film thermocouples (TFTCs) are widely used in high-temperature environment measurements and have the advantages of good stability and high thermoelectric voltage. However, different annealing processes affect the performance of TFTCs. This paper studied the impact of different annealing times on the cyclic characteristics of ceramic oxide thin film thermocouples. ITO/In_2O_3 TFTCs were prepared on alumina ceramics by a screen printing method, and the samples were annealed at different times. The microstructure of the ITO film was studied by scanning electron microscopy (SEM), X-ray diffraction (XRD), and X-ray photoelectron spectroscopy (XPS). The results show that when the annealing temperature is fixed, the stability of the thermocouple is worst when it is annealed for 2 h. Extending the annealing time can improve the properties of the film, increase the density, slow down oxidation, and enhance the thermal stability of the thermocouple. The thermal cycle test results show that the sample can reach five temperature rise and fall cycles, more than 50 h, and can meet the needs of stable measurement in high temperature and harsh environments.

Keywords: thin film thermocouples; temperature; anneal

Citation: Han, Y.; Ruan, Y.; Xue, M.; Wu, Y.; Shi, M.; Song, Z.; Zhou, Y.; Teng, J. Effect of Annealing Time on the Cyclic Characteristics of Ceramic Oxide Thin Film Thermocouples. *Micromachines* **2022**, *13*, 1970. https://doi.org/10.3390/mi13111970

Academic Editor: Hugo Aguas

Received: 19 October 2022
Accepted: 11 November 2022
Published: 13 November 2022

Publisher's Note: MDPI stays neutral with regard to jurisdictional claims in published maps and institutional affiliations.

1. Introduction

With the development of microelectromechanical system (MEMS) technology, the technology of using high-temperature sensors to measure the temperature of target pipelines has emerged [1–6]. For example, the working temperature of large-scale equipment such as iron and steel smelting, casting high temperature, and gas turbines is generally 600–1100 °C, of which the temperature of the exhaust flue is about 25–800 °C [7–9]. Monitoring the high temperature generated by the exhaust flue can judge the process production status, which is conducive to ensuring safe and effective production and research. In traditional technology, infrared radiation temperature measurement has been adopted, but the infrared temperature measurement requires that there is no barrier between the infrared probe and the surface to be measured of the exhaust gas flue, and that the surface to be measured is required to be free of contaminants and other impurities. However, it is difficult to ensure that the surface to be measured is free of pollutants in the actual industrial environment, which leads to low accuracy of the temperature of the monitored target pipeline.

Thermocouples can be classified into two categories: metal type and ceramic type. The traditional K-type and S-type thin film thermocouples do not meet the requirements of such a high-temperature test because they are easy to oxidize and the change of microstructure reduces their temperature measurement accuracy and life span. The S-type (Pt/Pt-10% Rh) thin film thermocouple is expensive, and the thermoelectric potential of the positive element decreases when working at 600~800 °C [10]. Metal thermocouples such as the Pt/Pd film thermocouple, with the progress of the thermal cycle, the microstructure changes such as the growth of pores and grains in the film lead to the decrease of thermoelectric potential, and the destruction of the electrode film structure will lead to the failure of the film thermocouple. Therefore, the stability of the film structure plays a crucial role in preventing the failure of the thin film thermocouple during the thermal cycling process [11]. On the contrary, thermocouples made of In_2O_3 and ITO (In_2O_3:SnO_2 = 90:10 wt%) show higher thermoelectric voltage and better thermal stability [4,12–15]. The annealing process will also affect the performance of the ITO/In_2O_3 thin film thermocouple. When annealing under N^2 at 500 °C, the thermoelectric response will be unstable when working at high temperatures for a long time, while the thermocouple annealed in air at 1000 °C shows better stability [16]. ITO is a typical oxide semiconductor, and annealing has a significant effect on its thermoelectric properties, especially the Seebeck coefficient and thermoelectric stability [7]. In the process of preparing ITO/In_2O_3 thin film thermocouples (TFTCs), it is found that the test results of the thermoelectric cycle cannot meet the needs. The linearity of the relationship curve between temperature and voltage is poor, but the linearity and repeatability are improved after several cycles; this phenomenon will make the sensor temperature test inaccurate and limit its application. We want to improve this phenomenon by improving the annealing process so that ITO/In_2O_3 TFTCs have a stable performance in the test process.

In this paper, ITO/In_2O_3 TFTCs were prepared on alumina ceramics by screen printing [17,18]. In contrast, ITO TFTCs annealed for different annealing times were studied accordingly. Different samples are made at 1000 °C for 1 h, 2 h, 3 h, and 5 h, respectively. The microstructure of the ITO film was studied by scanning electron microscopy (SEM), X-ray diffraction (XRD), and X-ray photoelectron spectroscopy (XPS). The results show that when the annealing temperature is fixed, the stability of the thermocouple is worst when it is annealed for 2 h. When the annealing time is above 3 h, the thermocouple can maintain stability and excellent linearity. The experimental phenomenon shows that extending the annealing time can improve the properties of the film, increase the density, slow down oxidation, and enhance the thermal stability of the thermocouple. The thermal cycle test can reach five temperature rise and fall cycles, and the time exceeds 50 h. Extending the high-temperature annealing time is conducive to the stability of the thermoelectric output of the thermocouple.

2. Materials and Methods

2.1. Fabrication of ITO/In_2O_3 Thin Film Thermocouple

ITO/In_2O_3 TFTCs were prepared by a screen printing process. The substrate is 99 alumina ceramic sheets (150 mm × 15 mm × 1 mm). The alumina ceramic sheet was fixed under the graphical screen printing plate (20 cm × 30 cm), which is 300 meshes; each electrode is 30–35 cm long and 0.2 cm wide. The scraper is made of rubber with a width of 5 cm. The fabrication process of ITO/In_2O_3 TFTCs by screen-printing is shown in Figure 1a. When printing, first of all, the position of the scraper needs to be adjusted to keep the scraper at a 45° angle with the printing plane, apply pressure to the screen printing plate with constant force and uniform speed, and squeeze the prepared ITO slurry onto the alumina ceramic substrate [19]. The ITO electrode is solidified after being heated at 100 °C for an hour in a furnace; then the In_2O_3 slurry is printed on the substrate by the same steps. The positive electrode and negative electrode of the thermocouple are completed. After being annealed at 800 °C for an hour, the stress in the film is released. Figure 1b shows the optical image of the ITO/In_2O_3 thermocouple. ITO/In_2O_3 thin film was then annealed in air at 1000 °C at different times. Different samples are made at 1000 °C for 1 h, 2 h, 3 h, and 5 h, respectively. ITO/In_2O_3 TFTCs were made. A copper wire (20 cm) was coated on the cold junction of ITO and In_2O_3 with silver paste and dried at 150 °C for 2 h.

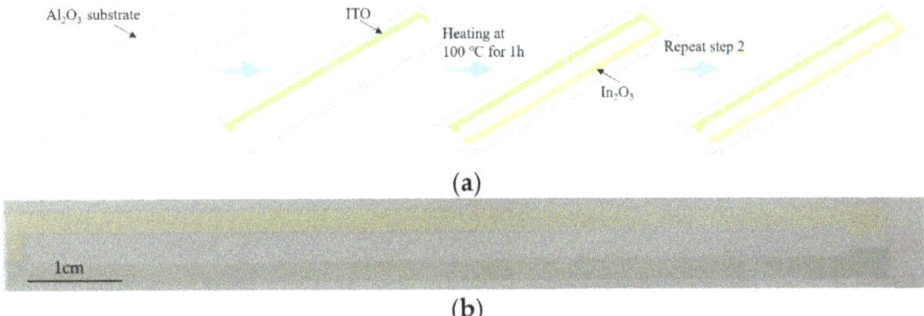

Figure 1. (**a**) Fabrication process of ITO/In$_2$O$_3$ TFTCs by screen-printing. (**b**) Optical image of the ITO/In$_2$O$_3$ thermocouple.

2.2. Measurements

Surface images were obtained by SEM. Using XRD, we analyzed the crystal structure of ITO and In$_2$O$_3$ films; the step was 0.02°. The chemical composition and chemical state were analyzed by XPS.

ITO/In$_2$O$_3$ TFTCs were statically calibrated with a standard calibration furnace. The standard S-type thermocouple is used to continuously monitor the furnace temperature, that is, the hot junction temperature and the platinum resistance is used to measure the cold junction temperature. ITO/In$_2$O$_3$ TFTCs were calibrated with the temperature of hot junction up to 1050 °C.

3. Results

Microstructures of ITO and In$_2$O$_3$ Films

SEM micrographs of ITO and In$_2$O$_3$ films treated for different annealing times are shown in Figure 2.

Figure 2. Surface morphology of the ITO ((**a1**) as-deposited; (**a2**) 1 h; (**a3**) 2 h; (**a4**) 3 h; (**a5**) 5 h), and In$_2$O$_3$ ((**b1**): as-deposited; (**b2**) 1 h; (**b3**) 2 h; (**b4**) 3 h; (**b5**) 5 h) films annealed for different annealing times.

The grain size of ITO and In$_2$O$_3$ films increases with the prolongation of heat treatment time. As can be seen from Figure 3, the surface morphology of the films changed slightly with the extension of heat treatment time, and all the films showed a similar density. It can be seen from Figure 4 that the In$_2$O$_3$ thin films are all cubic particles after annealing at different times, that is, the cubic shapes of In$_2$O$_3$. The grain-sized In$_2$O$_3$ films become denser with the prolongation of heat treatment time. This is because the more oxygen vacancies are occupied in the In_2O_3 film, the more sufficient the oxidation is. With the prolongation of annealing time, the space available for oxygen gradually decreases. Therefore, with the prolongation of heat treatment time, the film density changes slowly [20,21].

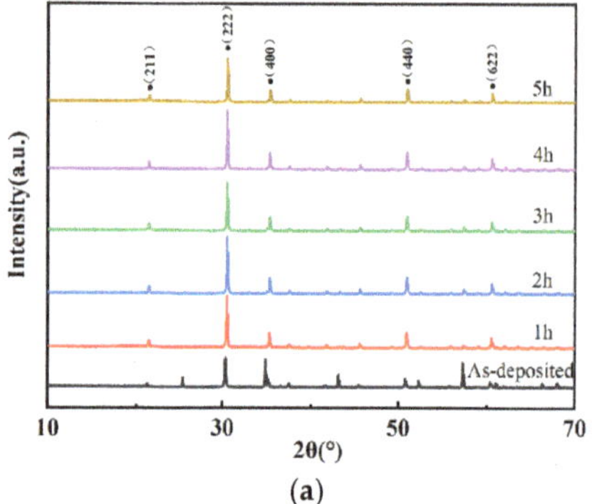

(a)

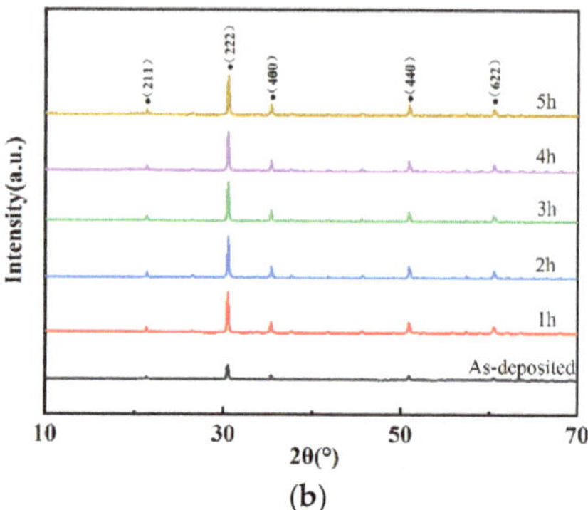

(b)

Figure 3. XRD patterns of films annealed for different annealing times: (a) ITO; (b) In_2O_3.

Figure 3 shows XRD patterns of ITO and In_2O_3 thin films for different annealing times. Figure 3a shows the as-deposited ITO film was amorphous. After being annealed at 1000 °C, there are diffraction peaks at 21.4°, 30.5°, 35.4°, 37.6°, 45.6°, 51.0°, and 60.6°, which correspond to (211), (222), (400), (440), and (622) of cubic bixbyite structure In_2O_3. The preferred growth direction of all ITO films is the (400) [22,23]. With the increase of annealing time, the diffraction peak corresponding to the (222) crystal plane of In_2O_3 shifts to a low angle. This is because high-temperature calcination leads to lattice expansion and an increase in crystal plane spacing, which makes the diffraction angle shift [24]. The crystal phase characteristic peaks of Sn are not shown in the figure, which means that Sn atoms replace the In atoms in the In_2O_3 crystal lattice [25–28], resulting in better sensitivity of the sensor [29].

The surface composition and elemental oxidation states of ITO thin film were investigated by XPS analysis. Figure 4a is the XPS full spectrum of the ITO film heat-treated at 1000 °C for 5 h, and the nuclear energy levels of In3d, Sn3d, and O1s were analyzed. Figure 4b shows that the peak of In $3d_{5/2}$ at 444.5 eV indicates the incorporation of In3+ in In_2O_3. As shown in Figure 4c, the two peaks at 486.5 eV and 495.0 eV were ascribed to Sn $3d_{5/2}$ and Sn $3d_{3/2}$, respectively, which are characteristic of Sn4+. The peak at 486.4 eV belongs to the binding energy of Sn $3d_{5/2}$, which corresponds to the Sn4+ bonding state in the Sn4+-bonded ITO lattice. Figure 4d shows that there were two peaks in the energy level of O, and the peak fitting was performed on the as-deposited and annealed for 5 h, as shown in Figure 4e,f. It can be seen from Figure 4f that O1s had two binding energy peaks, 530.0 and 531.2 eV, respectively, which corresponded to oxygen in the In-O bond and oxygen in the Sn-O bond [30].

Using the semi-quantitative fitting method, it was found that the atomic ratio of In_2O_3/SnO_2 was 9:1. XPS analysis showed the bonding state of In3+ and Sn4+ in the ITO film Figure 5 shows the XPS results of ITO films prepared at different annealing times. Compared with the as-deposited films, the oxygen atom ratio decreases after annealing at 1000 °C. With the increase of annealing time, the atomic ratios of O, In and Sn remain relatively stable.

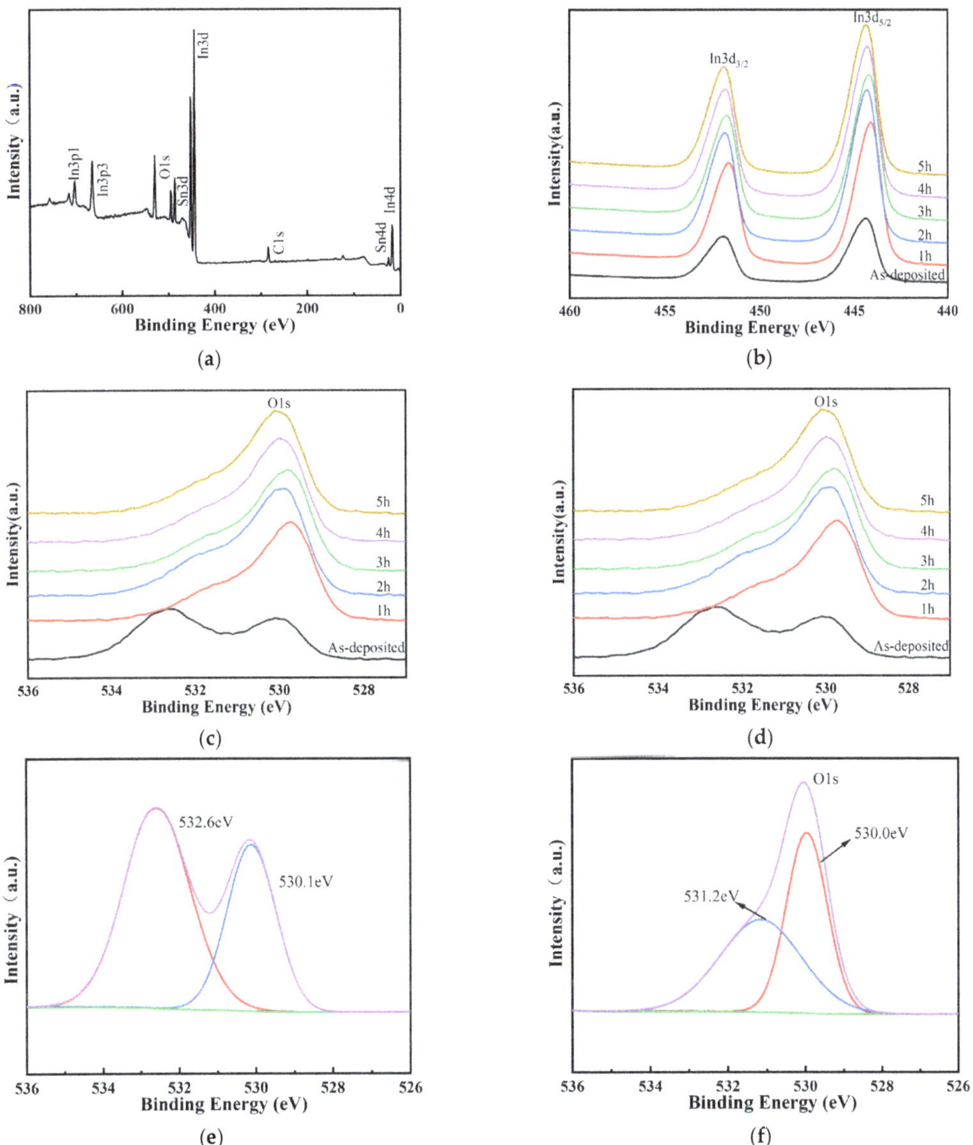

Figure 4. XPS of the ITO thin film: (**a**) full-spectra (**b–d**) XPS high-resolution peaks of In 3d, Sn 3d, and O1s, respectively; (**e**,**f**) XPS high-resolution peaks of O1s of as-deposited and 1 h.

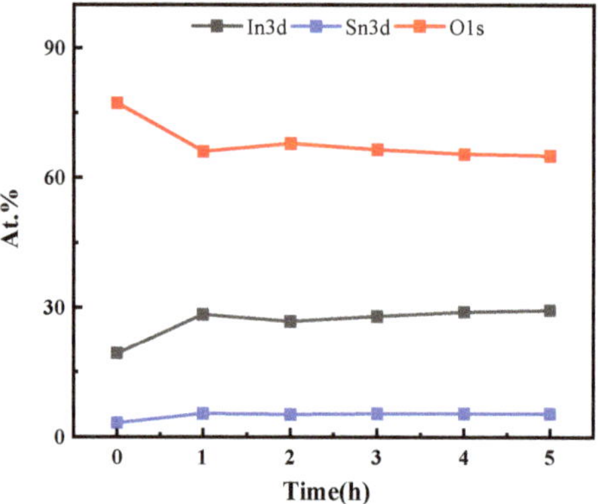

Figure 5. XPS results of ITO films fabricated for different annealing times.

Table 1 lists the conductivity of ITO and In_2O_3 films after annealing at different times. The conductivity increased with the prolongation of heat treatment time, which made the scattering of charge carriers decrease with the increase in grain size [15]. When the ITO film was annealed at 1000 °C for 5 h, its conductivity reached 3.82×10^3 S/m, which was 1.59 times that of the as-deposited sample. Therefore, the electrical conductivity of ITO films will be affected by the annealing time.

Table 1. The conductivity of ITO and In_2O_3 annealed at different times.

	Annealing Time (h)	0	1	2	3	5
Conductivity	ITO	2.41×10^3	2.9×10^3	3.25×10^3	3.48×10^3	3.82×10^3
(S/m)	In_2O_3	12.2	15.5	17.3	23.9	52.9

Figure 6 shows the thermoelectric output of the ITO/In_2O_3 TFTC from room temperature to 1150 °C. The thermal output of the thermocouple is stable over five thermal cycles and can be used over and over again. The stability of the thermocouple is the worst when it is annealed for 2 h in Figure 6b. As shown in Figure 6, ITO/In_2O_3 TFTC continues to work after five temperature rise and fall cycles, about 50 h, and the linearity is still high.

A third term polynomial is used to better describe the thermoelectric potential versus temperature in Equation (1):

$$V(\Delta T) = A(\Delta T)^3 + B(\Delta T)^2 + C(\Delta T) + D, \tag{1}$$

the fitting results are listed in Table 2. The fitting correlation coefficients (R^2) of all calibration curves are greater than 0.999. The coefficients of the third and second polynomials are smaller, indicating that the curve is nearly linear. According to the fitting results, the average Seebeck coefficient of ITO/In_2O_3 TFTCs that were annealed for five hours is 148.62 μV/°C. At the same time, the Seebeck coefficient of the thermocouple decreases with the annealing time of 3 h, which is attributed to the increase in carrier concentration [7]. We used the data in Table 2 as the standard temperature measurement expression of four groups of thin-film thermocouples to calculate the temperature deviation between the third calibration result and the expression. The result is shown in Figure 7; the expression of error is where ΔT is the temperature difference and $\Delta T1$ is the temperature difference value calculated by the formula in Table 2. The temperature measurement errors of the unannealed TFTCs reached 2.69%, which is very undesirable. However, the maximum temperature measurement errors of films annealed for 1 h, 2 h, 3 h, and 5 h are 0.58%, 1.39%, 0.70%, and 0.41%, respectively, and the temperature measurement errors of annealed for 5 h are not more than 0.50%.

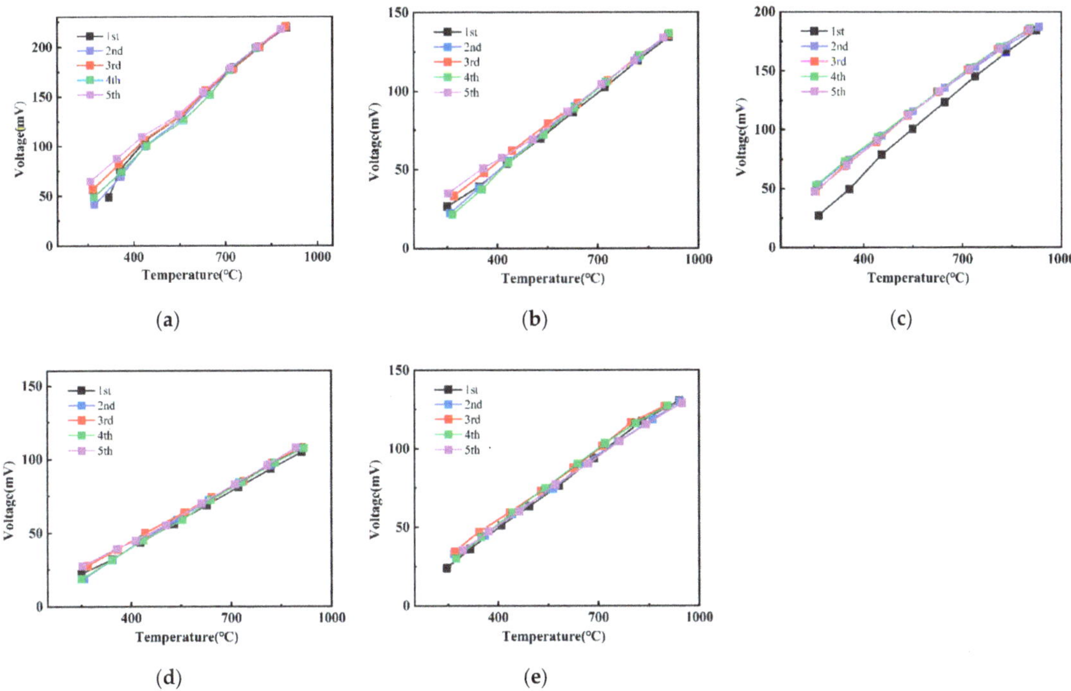

Figure 6. Thermoelectric output for different annealing times: (**a**) as-deposited; (**b**) 1 h; (**c**) 2 h; (**d**) 3 h; (**e**) 5 h.

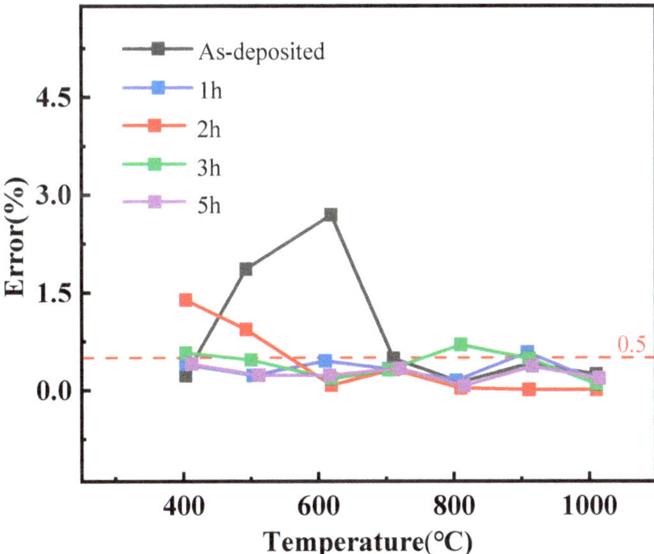

Figure 7. Errors of ITO/In$_2$O$_3$ TFTCs annealed for different times.

Table 2. Thermoelectric response and fitting results of ITO/In$_2$O$_3$ TFTCs for different annealing times.

Annealing Time (h)	$V(\Delta T) = A(\Delta T)^3 + B(\Delta T)^2 + C(\Delta T) + D$				Correlation Coefficient (R^2)	Average Seebeck Coefficient (μV/$^\circ$C)
	A (mV/$^\circ$C^3)	B (mV/$^\circ$C^2)	C (mV/$^\circ$C)	D (mV)		
0	7.52×10^{-8}	-1.54×10^{-5}	0.36	-13.04	0.99914	256.65
1	1.70×10^{-8}	-3.22×10^{-5}	0.18	-13.04	0.99981	155.29
2	-7.27×10^{-8}	7.17×10^{-5}	0.21	-8.89	0.99979	209.45
3	4.21×10^{-8}	-7.14×10^{-5}	0.16	-11.73	0.99979	123.75
5	-5.75×10^{-8}	5.80×10^{-5}	0.15	-14.27	0.99993	148.62

4. Conclusions

ITO/In$_2$O$_3$ TFTCs were prepared on alumina ceramics by screen printing. TFTCs annealed at different times were studied. The results show that when the annealing temperature is fixed, the stability of the thermocouple is the worst when it is annealed for 2 h. Extending the annealing time can improve the properties of the film, increase the density, slow down oxidation, and enhance the thermal stability of the thermocouple. The thermal cycle test results show that the sample can reach five temperature rise and fall cycles, more than 50 h.

Author Contributions: Conceptualization, Y.H., Y.R. and Y.W.; methodology, Y.H. and Y.W.; validation, Y.R., Y.W. and J.T.; formal analysis, Y.H.; investigation, Y.H. and M.X.; resources, M.S., Z.S. and Y.Z.; data curation, Y.H.; writing—original draft preparation, Y.H.; writing—review and editing, Y.R., Y.W. and J.T.; supervision, Y.R., Y.W. and J.T.; project administration, Y.R. and J.T.; funding acquisition, Y.R., Y.W. and M.S. All authors have read and agreed to the published version of the manuscript.

Funding: This research was funded by The National Key Research and Development Program of China (2020YFB2009103 and 2021YFB2011800).

Data Availability Statement: Not applicable.

Acknowledgments: The authors acknowledge Yong Zhu of AUS Griffith University for providing revision guidance. The authors acknowledge the MEMS Institute of Zibo National High-tech Industrial Development Zone for providing technical support.

Conflicts of Interest: The authors declare no conflict of interest.

References

1. Gregory, O.J.; You, T. Ceramic Temperature Sensors for Harsh Environments. *IEEE Sens. J.* **2005**, *5*, 833–838. [CrossRef]
2. Chen, Y.; Jiang, H.; Zhao, W.; Zhang, W.; Liu, X.; Jiang, S. Fabrication and Calibration of Pt–10%Rh/Pt Thin Film Thermocouples. *Measurement* **2014**, *48*, 248–251. [CrossRef]
3. Cougnon, F.G.; Depla, D. The Seebeck Coefficient of Sputter Deposited Metallic Thin Films: The Role of Process Conditions. *Coatings* **2019**, *9*, 299. [CrossRef]
4. Jin, X.H.; Ma, B.H.; Qiu, T.; Deng, J.J. ITO Thin Film Thermocouple for Transient High Temperature Measurement in Scramjet Combustor. In Proceedings of the 2017 19th International Conference on Solid-State Sensors, Actuators and Microsystems (Transducers), Kaohsiung, Taiwan, 18–22 June 2017; pp. 1148–1151.
5. Shi, P.; Wang, W.C.; Liu, D.; Zhang, J.; Ren, W.; Tian, B.; Zhang, J.Z. Structural and Electrical Properties of Flexible ITO/In$_2$O$_3$ Thermocouples on Pi Substrates under Tensile Stretching. *ACS Appl. Electron. Mater.* **2019**, *1*, 1105–1111. [CrossRef]
6. Wrbanek, J.; Fralick, G.; Farmer, S.; Sayir, A.; Blaha, C.; Gonzalez, J. Development of Thin Film Ceramic Thermocouples for High Temperature Environments. In Proceedings of the 40th AIAA/ASME/SAE/ASEE Joint Propulsion Conference and Exhibit, Fort Lauderdale, FL, USA, 11–14 July 2004.
7. Jin, X.H.; Ma, B.H.; Zhao, K.L.; Zhang, Z.X.; Deng, J.J.; Luo, J.; Yuan, W.Z. Effect of Annealing on the Thermoelectricity of Indium Tin Oxide Thin Film Thermocouples. *Ceram. Int.* **2020**, *46*, 4602–4609. [CrossRef]
8. Tougas, I.M.; Amani, M.; Gregory, O.J. Metallic and Ceramic Thin Film Thermocouples for Gas Turbine Engines. *Sensors* **2013**, *13*, 15324–15347. [CrossRef]
9. Meredith, R.D.; Wrbanek, J.D.; Fralick, G.C.; Greer, L.C.; Hunter, G.W.; Chen, L. Design and Operation of a Fast, Thin-Film Thermocouple Probe on a Turbine Engine. In Proceedings of the 50th AIAA/ASME/SAE/ASEE Joint Propulsion Conference 2014, Cleveland, OH, USA, 28–30 July 2014.

10. Kreider, K.G. Sputtered High Temperature Thin Film Thermocouples. *J. Vac. Sci. Technol. A Vac. Surf. Films* **1993**, *11*, 1401–1405. [CrossRef]
11. Tougas, I.M.; Gregory, O.J. Thin Film Platinum–Palladium Thermocouples for Gas Turbine Engine Applications. *Thin Solid Films* **2013**, *539*, 345–349. [CrossRef]
12. Gregory, O.J.; Busch, E.; Fralick, G.C.; Chen, X. Preparation and Characterization of Ceramic Thin Film Thermocouples. *Thin Solid Films* **2010**, *518*, 6093–6098. [CrossRef]
13. Wrbanek, J.D.; Fralick, G.C.; Zhu, D. Ceramic Thin Film Thermocouples for Sic-Based Ceramic Matrix Composites. *Thin Solid Films* **2012**, *520*, 5801–5806. [CrossRef]
14. Gregory, O.J.; Amani, M.; Tougas, I.M.; Drehman, A.J. Stability and Microstructure of Indium Tin Oxynitride Thin Films. *J. Am. Ceram. Soc.* **2011**, *95*, 705–710. [CrossRef]
15. Zhang, J.Z.; Wang, W.C.; Liu, D.; Zhang, Y.; Shi, P. Structural and Electric Response of ITO/In_2O_3 Transparent Thin Film Thermocouples Derived from Rf Sputtering at Room Temperature. *J. Mater. Sci. Mater. Electron.* **2018**, *29*, 20253–20259. [CrossRef]
16. Chen, X.; Gregory, O.J.; Amani, M. Thin-Film Thermocouples Based on the System In_2O_3-SnO_2. *J. Am. Ceram. Soc.* **2011**, *94*, 854–860. [CrossRef]
17. Tan, Q.L.; Lv, W.; Ji, Y.H.; Song, R.J.; Lu, F.; Dong, H.L.; Zhang, W.D.; Xiong, J.J. A Lc Wireless Passive Temperature-Pressure-Humidity (Tph) Sensor Integrated on Ltcc Ceramic for Harsh Monitoring. *Sens. Actuators B-Chem.* **2018**, *270*, 433–442. [CrossRef]
18. Zhang, T.; Tan, Q.L.; Lyu, W.; Lu, X.; Xiong, J.J. Design and Fabrication of a Thick Film Heat Flux Sensor for Ultra-High Temperature Environment. *IEEE Access* **2019**, *7*, 180771–180778. [CrossRef]
19. Liu, D.; Shi, P.; Ren, W.; Liu, Y.T.; Niu, G.; Liu, M.; Zhang, N.; Tian, B.; Jing, W.X.; Jiang, Z.D.; et al. A New Kind of Thermocouple Made of P-Type and N-Type Semi-Conductive Oxides with Giant Thermoelectric Voltage for High Temperature Sensing. *J. Mater. Chem. C* **2018**, *6*, 3206–3211. [CrossRef]
20. Hsu, C.H.; Geng, X.P.; Wu, W.Y.; Zhao, M.J.; Zhang, X.Y.; Huang, P.H.; Lien, S.Y. Air Annealing Effect on Oxygen Vacancy Defects in Al-Doped Zno Films Grown by High-Speed Atmospheric Atomic Layer Deposition. *Molecules* **2020**, *25*, 5043. [CrossRef]
21. Xu, H.Y.; Huang, Y.H.; Liu, S.; Xu, K.W.; Ma, F.; Chu, P.K. Effects of Annealing Ambient on Oxygen Vacancies and Phase Transition Temperature of VO_2 Thin Films. *RSC Adv.* **2016**, *6*, 79383–79388. [CrossRef]
22. Liu, Y.; Ren, W.; Shi, P.; Liu, D.; Zhang, Y.; Liu, M.; Ye, Z.G.; Jing, W.; Tian, B.; Jiang, Z. A Highly Thermostable In_2O_3/ITO Thin Film Thermocouple Prepared Via Screen Printing for High Temperature Measurements. *Sensors* **2018**, *18*, 958. [CrossRef]
23. Thilakan, P.; Kumar, J. Studies on the Preferred Orientation Changes and Its Influenced Properties on ITO Thin Films. *Vacuum* **1997**, *48*, 463–466. [CrossRef]
24. Ishigaki, N.; Kuwata, N.; Dorai, A.; Nakamura, T.; Amezawa, K.; Kawamura, J. Effect of Post-Deposition Annealing in Oxygen Atmosphere on Licomno4 Thin Films for 5 V Lithium Batteries. *Thin Solid Films* **2019**, *686*, 137433. [CrossRef]
25. Zhao, X.; Yang, K.; Wang, Y.; Chen, Y.; Jiang, H. Stability and Thermoelectric Properties of ITON:Pt Thin Film Thermocouples. *J. Mater. Sci. Mater. Electron.* **2015**, *27*, 1725–1729. [CrossRef]
26. Zhang, Y.; Cheng, P.; Yu, K.Q.; Zhao, X.L.; Ding, G.F. ITO Film Prepared by Ion Beam Sputtering and Its Application in High-Temperature Thermocouple. *Vacuum* **2017**, *146*, 31–34. [CrossRef]
27. Debataraja, A.; Zulhendri, D.W.; Yuliarto, B.; Sunendar, B. Investigation of Nanostructured Sno 2 Synthesized with Polyol Technique for Co Gas Sensor Applications. *Procedia Eng.* **2017**, *170*, 60–64. [CrossRef]
28. Hemasiri, B.W.N.H.; Kim, J.K.; Lee, J.M. Synthesis and Characterization of Graphene/ITO Nanoparticle Hybrid Transparent Conducting Electrode. *Nano-Micro Lett.* **2017**, *10*, 18. [CrossRef]
29. Yadav, B.C.; Agrahari, K.; Singh, S.; Yadav, T.P. Fabrication and Characterization of Nanostructured Indium Tin Oxide Film and Its Application as Humidity and Gas Sensors. *J. Mater. Sci. Mater. Electron.* **2016**, *27*, 4172–4179. [CrossRef]
30. Wu, H.; Gan, Z.; Chu, X.; Liang, S.; He, L. Preparation and Gas-Sensing Properties of One-Dimensional Ga_2O_3/SnO_2 Nanofibers. *Chin. J. Inorg. Chem.* **2020**, *36*, 309–316.

 *micromachines*

Article

Design and Optimization of Hemispherical Resonators Based on PSO-BP and NSGA-II

Jinghao Liu, Pinghua Li, Xuye Zhuang *, Yunlong Sheng *, Qi Qiao, Mingchen Lv, Zhongfeng Gao and Jialuo Liao

College of Mechanical Engineering, Shandong University of Technology, Zibo 255000, China; liuf162999@163.com (J.L.)
* Correspondence: zxye8888@163.com (X.Z.); shengyunlong@sdut.edu.cn (Y.S.)

Abstract: Although one of the poster children of high-performance MEMS (Micro Electro Mechanical Systems) gyroscopes, the MEMS hemispherical resonator gyroscope (HRG) is faced with the barrier of technical and process limits, which makes it unable to form a resonator with the best structure. How to obtain the best resonator under specific technical and process limits is a significant topic for us. In this paper, the optimization of a MEMS polysilicon hemispherical resonator, designed by patterns based on PSO-BP and NSGA-II, was introduced. Firstly, the geometric parameters that significantly contribute to the performance of the resonator were determined via a thermoelastic model and process characteristics. Variety regulation between its performance parameters and geometric characteristics was discovered preliminarily using finite element simulation under a specified range. Then, the mapping between performance parameters and structure parameters was determined and stored in the BP neural network, which was optimized via PSO. Finally, the structure parameters in a specific numerical range corresponding to the best performance were obtained via the selection, heredity, and variation of NSGAII. Additionally, it was demonstrated using commercial finite element soft analysis that the output of the NSGAII, which corresponded to the Q factor of 42,454 and frequency difference of 8539, was a better structure for the resonator (generated by polysilicon under this process within a selected range) than the original. Instead of experimental processing, this study provides an effective and economical alternative for the design and optimization of high-performance HRGs under specific technical and process limits.

Keywords: PSO; BP neural network; NSGA-II; MEMS HRG

Citation: Liu, J.; Li, P.; Zhuang, X.; Sheng, Y.; Qiao, Q.; Lv, M.; Gao, Z.; Liao, J. Design and Optimization of Hemispherical Resonators Based on PSO-BP and NSGA-II. *Micromachines* **2023**, *14*, 1054. https://doi.org/ 10.3390/mi14051054

Academic Editors: Yong Ruan, Weidong Wang and Zai-Fa Zhou

Received: 14 April 2023
Revised: 29 April 2023
Accepted: 3 May 2023
Published: 16 May 2023

1. Introduction

The MEMS hemispherical resonator gyroscope (HRG) is one of the typical representatives of micro-resonator gyroscopes for its high reliability, good accuracy, strong stability, long service life, and simple structure [1,2]. The existing MEMS hemispherical resonator gyroscopes have achieved bias stability and a Q factor to the level of $0.01°/$h and 1,000,000, respectively. To adapt it to a different application scenario, researchers have developed the micro HRG for different materials and structures, and for these different gyros, there are a few discrepancies in their manufacturing processes.

Up until now, there have been three main approaches to processing a MEMS hemispherical resonator. Since the approaches are different, the resonators have different modeling characteristics. Additionally, since the manufacturing processes are different, the resonators have different modeling characteristics [3].

Andrei M. Shkel et al. from the University of California proposed the method of using glass blow molding to make the resonator of a MEMS HRG [4]. K. Najafi et al. from the University of Michigan developed a MEMS HRG with a bias stability of $0.0391°/$h and an angular random walk of $0.0087°/\sqrt{}$ h by applying a constant velocity of airflow directly above the fused silica [5].

On the other hand, it is difficult to control the concentricity error and roundness error of a resonator processed via this method, as Shanghai Jiao Tong University demonstrated [6].

Based on the silicon substrate, our research group obtained a hemispherical cavity via wet etching and then matched it with a hemispherical shell resonator using a polysilicon thin film deposition process [7]. J. M. Gray et al. from the University of Colorado in the United States obtained a resonator, of which the thickness was as small as 50 nm based on ALD (atomic layer deposition) [8]. However, the dimensional parameters of a resonator processed through this method usually stay at an uncontrollably larger size, as the National University of Defense Technology has shown [9,10].

The Shanghai Institute of Optics and Fine Mechanics, the Chinese Academy of Sciences, and the National University of Defense Technology all developed a resonator successfully by laser etching, whose diameter was 39 nm and 1 mm, respectively [11,12].

Although there has been a flood of research to reveal the manufacturing process mechanisms to improve the modeling quality of the resonator, many problems remain [11,13], as shown in Table 1. In contrast to the macroscopic hemispherical gyroscope, whose shaping characteristics can be eliminated via accurate post-processing, the MEMS hemispherical gyroscope has obvious shaping characteristics in its forming structure that cannot be compensated for due to its excessively small size. It is known that a different gyro structure corresponds to a different processing method and that this cannot be chosen [14,15]. So, experience always plays an important role in the design of a resonator, which cannot be separated from experimental exhaustion and processing. On the one hand, it is inefficient to obtain all of the resonator structures through enumeration; on the other hand, it is impossible to obtain the optimal structure through enumeration without consideration for the characteristics corresponding to the existing manufacturing process. Thus, the question of how to obtain the optimal structure under the restriction of a specific manufacturing process using a systematic approach is one of the most important projects for micro-HRG.

Table 1. Characteristics of different manufacturing processes.

Manufacturing Process	Structure	Limits
Glass blow molding	Large birdbath with thin bottom	Complex process
Laser ablation	Relatively standard hemisphere	Complex assembly, Difficult to machine complex surface
Thin film deposition	Hemisphere with non-standard aspect ratio	Large surface roughness, Large frequency cracking

2. Data Collection and Analysis Based on PSO-BP and NSGA-II

As Figure 1 shows, MEMS HRG achieves its high-precision detection through the standing wave influenced by Coriolis affecting the resonator. The incentive part affects periodic excitation caused by coulomb forces or mechanical forces to the resonator; then, a standing wave will come into play on the resonator. The antinode points of the standing wave are set as A, B, C, and D; when the shell resonator is stimulated by an angular velocity Ω relative to the inertial system, each antinode point will have the implicated angular velocity Ω.

Under the effect of V_A, V_B, V_C, and V_D caused by the standing wave and the implicated angular velocity Ω external input of A, B, C, and D have Coriolis accelerations of W_{KA}, W_{KB}, W_{KC}, and W_{KD}, respectively. Thus, the Coriolis inertial force of P_{KA}, P_{KB}, P_{KC}, and P_{KD} are generated by the Coriolis accelerations of V_A, V_B, V_C, and V_D. Two equivalent Coriolis couples with opposite directions are formed via P_{KA}–P_{KC} and P_{KB}–P_{KD}, resulting in the precession of the standing wave relative to the resonator shell and inertial reference frame.

Due to the precession of the standing wave, the points on the harmonic oscillator that have an isometric deformation will change to other ones. Measuring electrical signal changes at specific points caused by their deformation changes, the accurate angle and angular velocity of the external input will be determined.

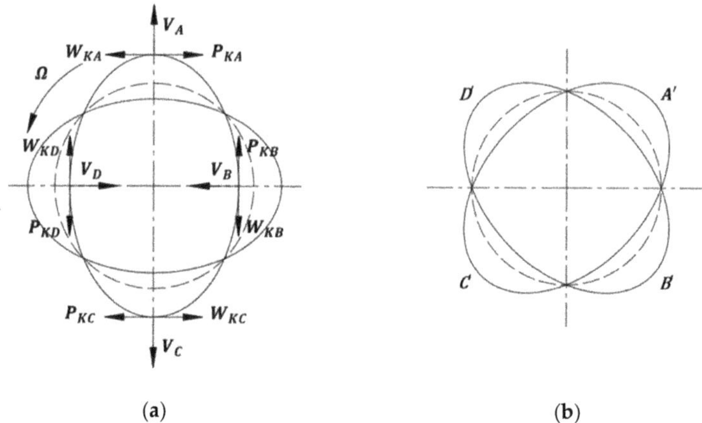

Figure 1. Operating mechanism of HRG: (**a**) drive mode of the HRG and (**b**) detect mode of the HRG.

The signal quality is affected by multiple factors, of which the most significant effect is the Q factor and frequency characteristic.

The Q factor represents the ratio of the total energy that the resonator stored to the energy that dissipated in a vibration cycle. There are a few elements that influence the Q factor, such as thermoelastic damping, anchor damping, surface damping, and air damping. The Q factor can be written as follows [16]:

$$\frac{1}{Q} = \frac{1}{Q_{surface}} + \frac{1}{Q_{ted}} + \frac{1}{Q_{air}} + \frac{1}{Q_{anchor}} \tag{1}$$

Air damping represents the energy consumption caused by air viscous force to a certain degree. However, since the MEMS resonator always works in a vacuum environment, it does not work obviously in energy consumption for the air damping.

The anchor damping represents the energy consumption caused by anchor loss. When the resonator vibrates, a part of the energy will be lost through the basement to the earth, and that is what is called anchor loss.

Surface damping represents the energy consumption caused by surface defects such as large surface harshness. A rough surface causes increasing surface quality imbalances and other problems, which make the resonator have to carve out a portion of energy to keep balance.

Thermal elastic damping represents the energy consumption caused by the irreversible heat flow. Under the effect of the standing wave, the shell resonator receives tension and compression at the different parts; then, the temperature of the different parts under tension and compression decreases and increases, respectively, resulting in a local temperature difference in the resonator.

For the MEMS HRG, the size and deformation of the anchor are too small to make the anchor loss give considerable influence on the Q factor compared with the thermal electric damping. Additionally, as shown above, surface damping and air damping can also be ignored. So, what we have to take into consideration is only thermal elastic damping.

The frequency characteristic includes characteristic frequency and frequency cracking. Characteristic frequency is the natural frequency of a specific vibration mode; it is influenced by the structure and materials of the resonator. Frequency cracking is the difference between the operating mode and its parasitic mode; it is influenced by structure mass imbalance and stiffness imbalance of the resonator.

The operating mode of HRG is the natural-bending vibration mode, also known as the four-antinode mode. Generally, the ambient noise is around 5000 Hz, aiming at minimizing the disturbance to measurement signal caused by ambient noise; the characteristic frequency of operating mode should keep a distance to the frequency of 5000 Hz.

2.1. Preparation

For this study, an ideal model was used, in which the stiffness and mass are absolutely symmetrical; thus, the frequency cracking was not taken into consideration.

Additionally, for the resonator to have other modes at low-frequency ranges except for the operating mode, it is necessary to make the difference between the operating mode and other low-frequency modes as large as possible.

To optimize the resonator based on the analysis of thermal elastic damping, the trajectories of the irreversible heat flow should be determined. In the heat conduction of the solid module and solid mechanics module of COMSOL, as the finite element simulation results revealed, the trajectories of the irreversible heat flows are shown in Figure 2.

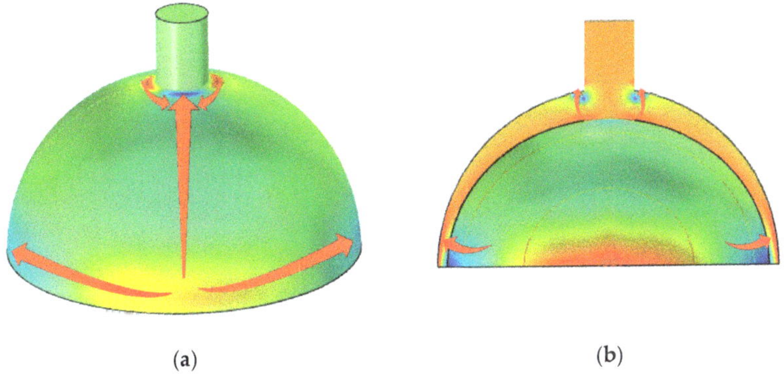

(a)   (b)

Figure 2. The trajectories of irreversible heat flow in operating mode: (**a**) heat flow on the surface and (**b**) heat flow on the cross-section.

There are five tracks for the irreversible heat flows (refer to [17]). The heat flows along the lip of the resonator and the anchor of the hemispherical shell surface are relatively long, and the heat transfer time is much longer than the vibration period of the operating mode. Therefore, the energy loss caused by these three heat flows can be ignored. Then, the thermo elastic damping of the resonator can be written as follows:

$$\frac{1}{Q_{\text{TED}}} = \frac{1}{Q_R} + \frac{1}{Q_r} \tag{2}$$

$$Q_R = \left(\frac{E\alpha^2 T_0}{C\rho}\right) \frac{\omega \cdot \tau_{\text{across}-t}}{1 + (\omega \cdot \tau_{\text{across}-t})^2} \tag{3}$$

$$\frac{1}{Q_r} = \left(\frac{C\rho}{E\alpha^2 T_0}\right) \frac{1 + (\omega \cdot \tau_{\text{across}-z})^2}{\omega \cdot \tau_{\text{across}-z}} + \left(\frac{C\rho}{E\alpha^2 T_0}\right) \frac{1 + (\omega \cdot \tau_{\text{across}-s})^2}{\omega \cdot \tau_{\text{across}-s}} \tag{4}$$

$$\tau_{\text{across}-t} = \frac{t^2 \rho C}{\pi^2 k} \tag{5}$$

$$\tau_{\text{across}-z} = \frac{\bar{z}^2 \rho C}{\pi^2 k} \tag{6}$$

$$\tau_{\text{across}-s} = \frac{s^2 \rho C}{4k} \tag{7}$$

ρ, C, E, α, and T_0 are the density, constant pressure heat capacity, Young's modulus, coefficient of thermal expansion, and initial temperature of the harmonic oscillator material, respectively; $\tau_{across-h}$, $\tau_{across-z}$, and $\tau_{across-s}$ are time constants for the heat flow transfer through the thickness direction of the hemispherical shell, the heat flow transfer through the thickness direction at the anchor, and the heat flow along the circumferential direction of the edge of the anchor, respectively; t is the thickness of the edge of the shell; $\bar{z}$ is the average length of the heat flow passing through the thickness direction at the anchor; s is the radius of the hemispherical shell at the edge of the anchor.

It can be inferred that structure parameters that have an influence on the thermo elastic damping are t, $\bar{z}$, and s from (5)–(7). Additionally, the fundamental parameters of the structure parameters are a, R, r, H, H − h, and R − r (refer to Figure 3).

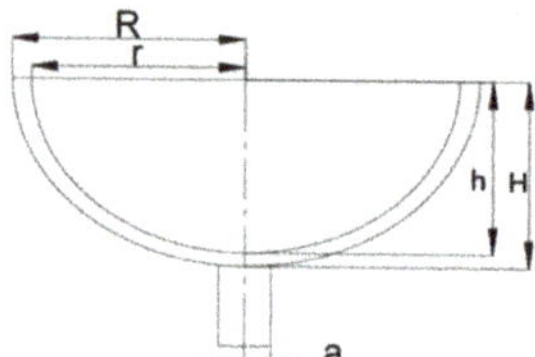

Figure 3. The structure-parameters of resonator.

Additionally, for this study, a polysilicon resonator processed via thin film deposition was taken as an example. The process of thin film deposition can be obtained from [18], as shown in Figure 4. First, the thermal oxide is patterned to mask the wafer, capacitive actuation, and sense electrodes that are isolated from the substrate are created by PN junctions. Second, A PECVD oxide layer is patterned to form the isotropic etch mask, and sulfur hexafluoride (SF6) plasma is used to etch the hemispherical mold. Third, an LPCVD low-stress silicon nitride layer is patterned to mask the topside and backside during KOH etching, which forms the back-side plug mold. Finally, the polysilicon layer on the wafer surface is removed in a short dry etching step with SF6 plasma while the shell is protected by a photoresist, and the shell is then released in HF.

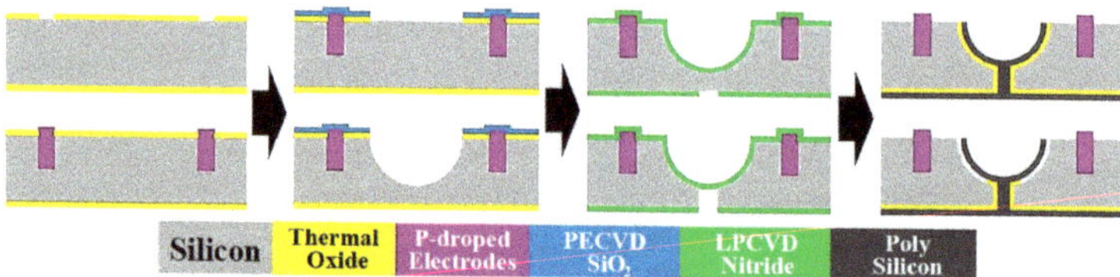

Figure 4. Process for preparation of resonator by thin film deposition.

There are mainly four steps for thin film deposition to process a resonator: first, manufacturing a mask on the silicon substrate; second, etching the silicon substrate to form a mold; third, depositing the structure in the mold; fourth, releasing the structure via wet etching.

The forming defects of the resonator processed via this method are mainly caused by the mold, which may cause a rough surface and a cavity structure, with the ratio of depth to width <1 resulting from wet etching, which cannot be resolved faultlessly now.

As mentioned earlier, the forming defects of specific manufacturing processes were the problems that should be put first. So, in consideration of the modeling characteristic and gravity effect when forming the structure via deposit, the parameter characteristics of the resonator can be written as follows:

$$\mathbf{AS} = \mathbf{b} \tag{8}$$

$$A = \begin{bmatrix} 1 & -1 & 0 & 0 & 0 \\ -1 & 1 & 0 & 0 & 0 \\ 1 & 0 & 0 & -1 & 0 \\ -1 & 0 & 0 & 1 & 0 \end{bmatrix} \tag{9}$$

$$S = \begin{bmatrix} R & r & H & h & a \end{bmatrix} \tag{10}$$

$$b = \begin{bmatrix} 2 \\ 0.5 \\ 50 \\ 50 \end{bmatrix} \tag{11}$$

A is the coefficient matrix; S is the structure parameters matrix; b is the deviation matrix. Set t as follows:

$$t = H - h \tag{12}$$

Referring to [18], the varying range of structure parameters was set as shown in Table 2, and polysilicon was selected to deposit the structure.

Table 2. Varying range of structure parameters.

R (μm)	r (μm)	H (μm)	h (μm)	A (μm)
301~502	299~500	251~552	249~550	50~100

Obtaining the mapping between structure parameters and performance parameters requires a lot of calculations; finite element simulation can solve the problem of data collection, and BP neural networks can solve the problem of fitting efficiently.

The variation is regular between performance parameters, and singular structure parameters can be obtained from simulation by varying one of the structure parameters and holding on to the others sequentially. Some of the results are shown in Table 3.

Table 3. Results of variable-controlling simulation.

R (μm)	r (μm)	H (μm)	h (μm)	a (μm)	Characteristic Frequency	Q Factor	Near Frequency
301	300	301	299.5	75	52,490.33	40,665.81	61,882.17
451	450	451	450	90	16,207.31	58,436.62	26,140.7
401	400	401	400	79	18,720.86	59,591.18	14,057.15
381.25	380	381	380	75	21,309.55	78,224.75	24,546.6
501	500	501	500	87	11,118.6	74,553.32	10,232.95

The curve graphs drawing the singular structure parameter with all of the performance parameters are shown in Figures 5–10. For every structure parameter, the figure (a–d) demonstrate when R or r is 300, 350, 400, and 500 as samples.

It can be seen that the effects of the performance parameters of varying R and r are almost the same, and it is in conformity with the conclusion that [17] provided. The increase in R and decrease in r both result in the increase in thickness of the edge of the shell, and it affects the path length of the heat flow transfer through the thickness direction of the hemispherical shell and the heat flow transfer through the thickness direction at the anchor.

Increasing the thickness of the hemispherical shell appropriately can extend the path length of the heat flow transfer through the thickness direction of the hemispherical shell and the heat flow transfer through the thickness direction at the anchor, decreasing the energy loss in a vibration period, which results in the decrease in thermoelastic damping, improve the Q factor, and raise the characteristic frequency.

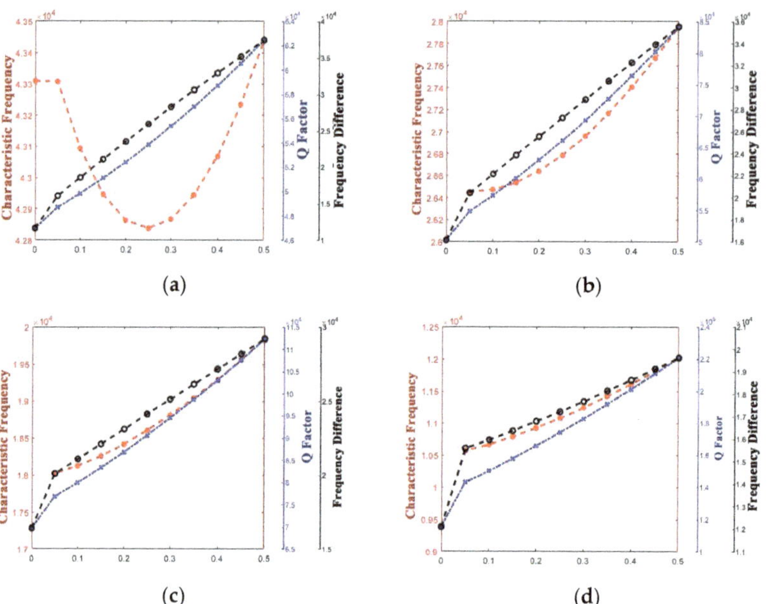

Figure 5. Variety regulation between R and performance parameters. (**a**) r = 300; (**b**) r = 400; (**c**) r = 450; (**d**) r = 500.

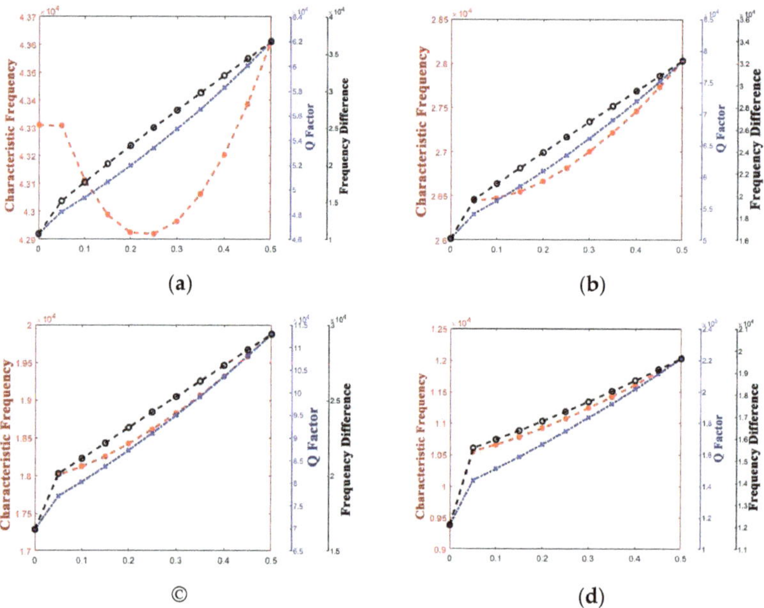

Figure 6. Variety regulation between r and performance-parameters (**a**) R = 300; (**b**) R = 400; (**c**) R = 450; (**d**) R = 500.

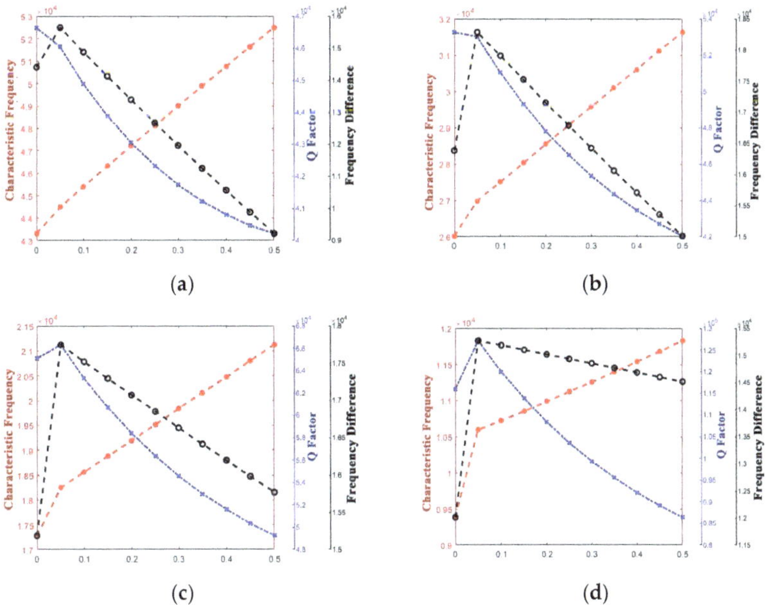

Figure 7. Variety regulation between t and performance-parameters (**a**) R = 300; (**b**) R = 400; (**c**) R = 450; (**d**) R = 500.

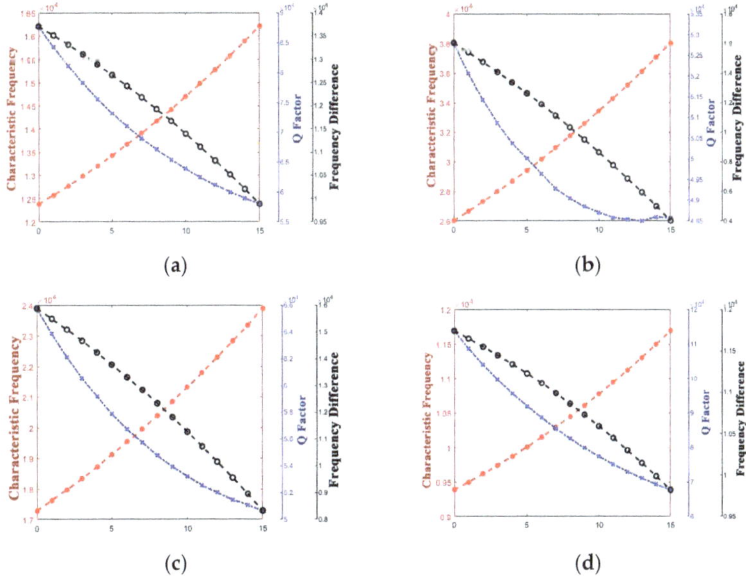

Figure 8. Variety regulation between a and performance parameters. (**a**) R = 300; (**b**) R = 400; (**c**) R = 450; (**d**) R = 500.

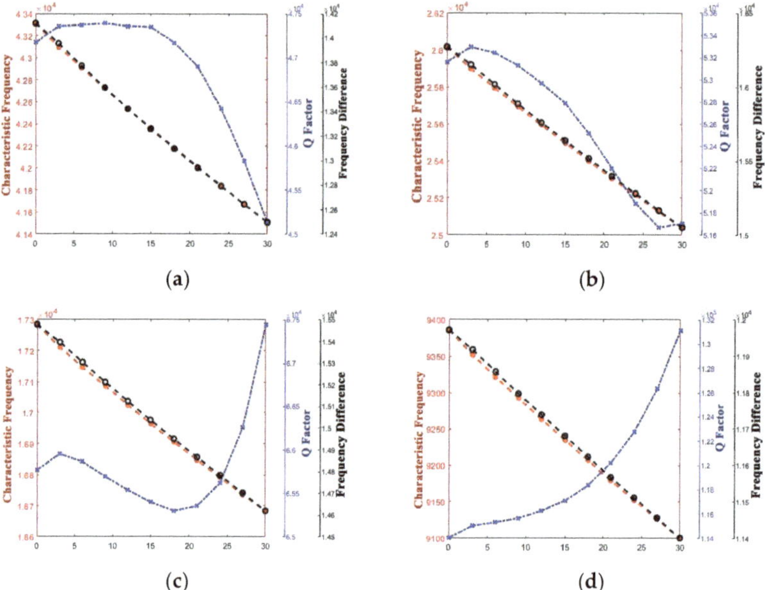

Figure 9. Variety regulation between h (increase) and performance parameters (**a**) R = 300; (**b**) R = 400; (**c**) R = 450; (**d**) R = 500.

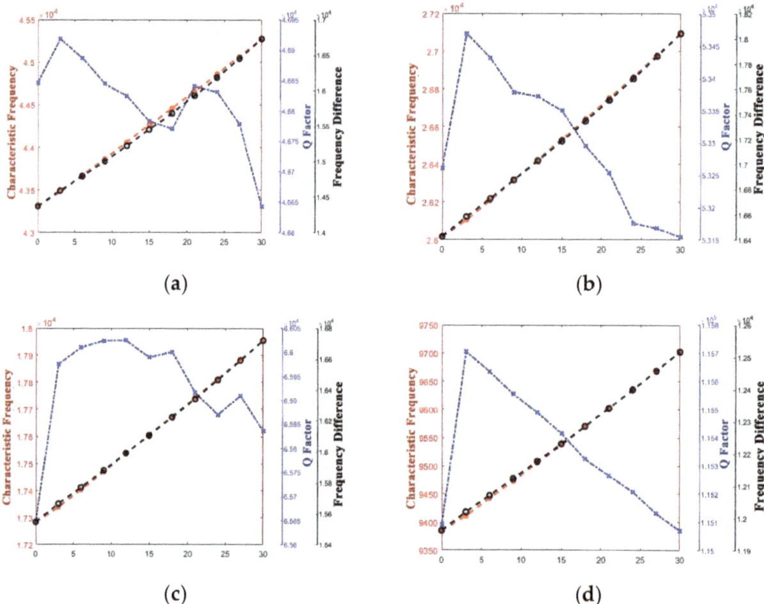

Figure 10. Variety regulation between h (decrease) and performance parameters. (**a**) R = 300; (**b**) R = 400; (**c**) R = 450; (**d**) R = 500.

However, whether it is the Q factor, the characteristic frequency, or the frequency difference, there is a turning point when the structure transfers from the standard hemisphere to the non-standard hemisphere. It can be inferred that it is the sudden change in the structure that results in the sudden change in the path length for the irreversible heat flow.

It is obvious that except for the turning point, there is a significant linear positive correlation between t and frequency difference and a linear negative correlation between t and characteristic frequency. It means that the characteristic frequency of the operating mode is getting closer and closer to the characteristic frequency of the other mode, which can also be said that the characteristic frequency is getting more and more intense. So, it is necessary for us to make the value of t as small as possible.

Additionally, for the Q factor, it is also necessary to control the size of t at a low level as it shows a negative correlation between t and the Q factor since the heat production converged from the edge to the bottom, which results in the increase in energy loss with the t increase.

It can be informed that the characteristic frequency increases as a increases. However, the Q factor and frequency difference are both generally negatively correlated with a; it can be inferred that although the path length of the heat flow at the anchor increased, the heat flow generated at the bottom also increased while a increased.

Figures 9 and 10 show the variety of regulation between performance parameters and h.

When H and h increase synchronously, out of our expectation, the Q factor decreases within a certain range when R is less than 500 μm, and then the Q factor increases, which is par for the course. An inference that the Q factor decrease because of the increase in energy loss caused by heat flow along the circumferential direction of the edge of the anchor is less than the decrease in energy loss caused by heat production converging from the edge to the bottom was given. Additionally, this phenomenon disappears when R is large enough.

When H and h decrease synchronously, predictably, the Q factor decreases as the energy loss caused by heat flow along the circumferential direction of the edge of the anchor decreases. Since the path length of heat flow along the circumferential direction of the edge of the anchor is reduced, this results in a reduced transfer time.

It can be seen that although the curve graphs can reflect the relationship between single structure parameter and performance parameters to some extent, it is difficult to explain the mapping reasonably since every structure parameters interact with each other, and it is, hence, the need for global optimization of all of the structure parameters.

2.2. Data Handling

BP neural network is one of the most widely used neural network models as a multi-layer feedforward neural network whose training strategy is according to an error back-propagation algorithm, and its structure is shown in Figure 11. We can obtain the mapping between multiple arguments and multiple variables via the BP neural network.

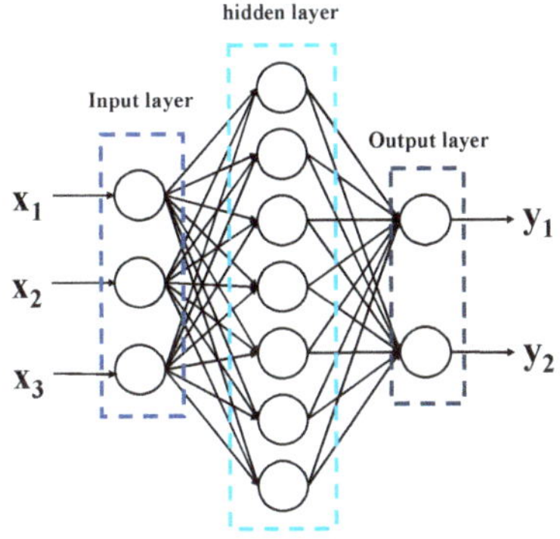

Figure 11. The structure of BP neural network.

However, the adjusted direction of weight values of traditional BP neural network is along the direction of local improvement, which leads to local extremum easily. Moreover, due to the BP neural networks being sensitive to initial weight values, different initial weight values may lead the neural network convergence to different local minima; in addition, there lacks an effective selection method to determine the weight values, which can only be obtained through experimental learning and training, while too many hidden layer neurons may lead to over learning, but too few causes under learning. Additionally, another problem is that the number of hidden layer neurons can only be roughly obtained based on experience, making the algorithm unstable.

Particle Swarm Optimization (PSO) can solve the above problems of traditional BP neural networks to some extent. We assume that the deviations, weight values, and a number of hidden layer neurons of the neural network as a disaggregation set to be solved, and all of the disaggregation sets are treated as a population in the PSO, while each disaggregation set is treated as an individual, as shown as Figure 12. This way, we can prevent the optimal weight values and deviation obtained from falling into the locally optimal solution by combining the locally optimal solution (individual optimal solution) and the globally optimal solution (population optimal solution) and obtain uniquely determined weight values, deviation, and a number of hidden layer neurons.

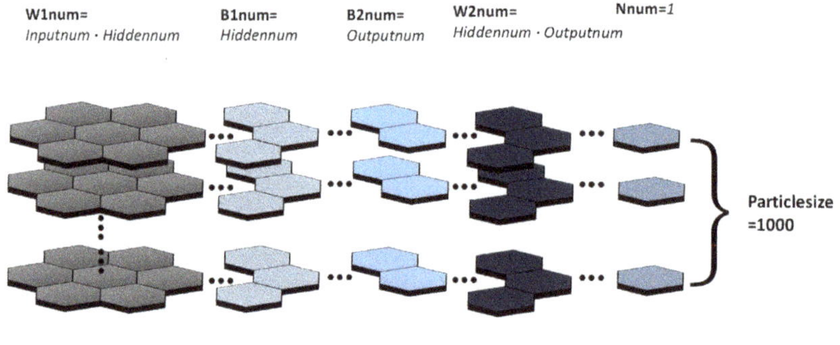

Figure 12. The structure of a particle for PSO-BP.

N is the number of hidden layers neurons; W1num, B1num, W2num, B2num, and N num correspond to the dimensions of W1, B1, W2, B2, and N. Input num is the dimension of input data, hidden num is the dimension of hidden layer neurons, output num is the dimension of output data, and N num is 1, which is used to record the number of hidden layer neurons.

NSGA-II is a kind of genetic algorithm which is always used to solve the combinatorial optimization problem; it can improve the probability of excellent individuals being preserved and reduce global complexity with an elitist strategy and the concept of congestion degree.

The mapping of structure parameters and performance parameters can be obtained via the BP neural network, and the BP neural network will be set as the constraint function of NSGA-II. Then, the best disaggregation set for the combination of structure parameters that correspond to the best performance parameters, under the limitation of a specific manufacturing processing, will come out as the output of NSGA-II, as shown in Figure 13.

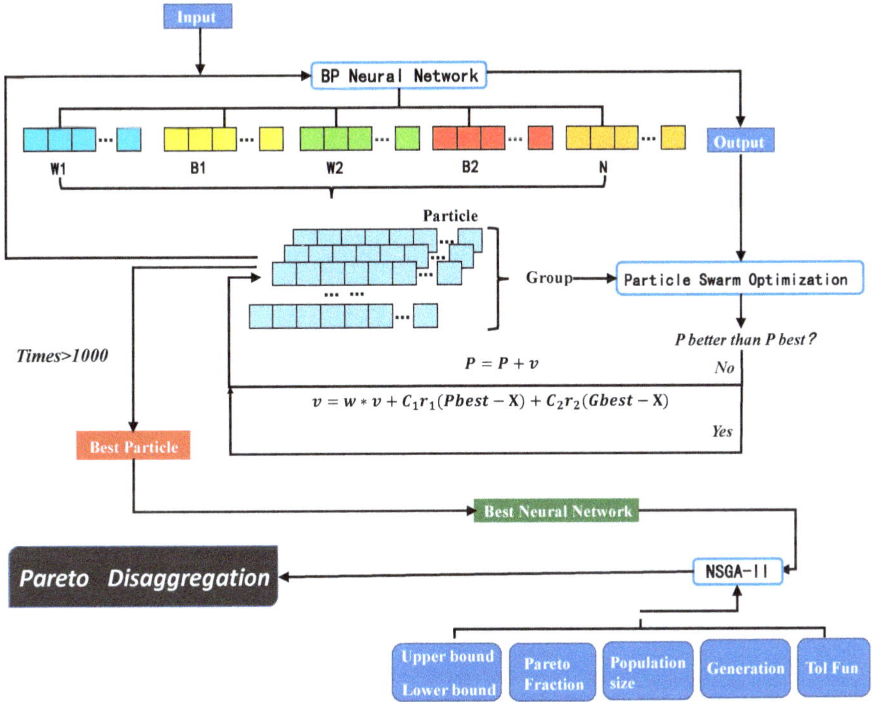

Figure 13. The structure of the algorithm for parameters optimization.

The specific steps are as follows:

1.   Train BP neural network.

    (1)   Build a sample set from simulation data. The dimension of the input is 5, and each input contains R, r, H, h, and a. The dimension of the output is 3, and each output contains f, f_0, and Q. There are 1760 sets of data. The matrix of input is as follows:

$$\mathbf{X} = [\mathrm{R\,r\,H\,h\,a}]^{\mathrm{T}} \tag{13}$$

The matrix of input is as follows:

$$\mathbf{Y} = \left[\mathrm{f\,f^{0}\,Q}\right]^{\mathrm{T}} \tag{14}$$

    (2)   Normalize the dataset as follows:

$$X_{i} = \frac{x_{i} - x_{\min}}{x_{\max} - x_{\min}}\,(i = 1, 2 \ldots \ldots 1760) \tag{15}$$

X_i is the normalized data of the No. i group, and x_i is the original data of the No. i group in the sample set. $x_{\max}$ and $x_{\min}$ are the original maximum and minimum in the sample set, respectively.

    (3)   Optimize the initial weight values, deviation, and number of hidden neurons of the BP neural network via PSO. The number of particles is set to 200, and the encoding of a single particle is as follows:

$$P1 = \left[w_{1,1}^{1}, \ldots\ldots, w_{N,5}^{1}, b_{1}^{1}, \ldots\ldots, b_{N}^{1}, w_{1,1}^{2}, \ldots\ldots, w_{3,N}^{2}, b_{1}^{2}, b_{2}^{2}, b_{3}^{2}, N\right] \tag{16}$$

$w_{1,1}^1, \ldots \ldots, w_{N,5}^1$ are the weight values from the input layer to the hidden layer, recorded as $\mathbf{W_1}$; $b_1^1, \ldots \ldots, b_N^1$ are the deviations from the input layer to the hidden layer, recorded as $\mathbf{B_1}$; $w_{1,1}^2, \ldots \ldots, w_{3,N}^2$ are the weight values from the hidden layer to the output layer, recorded as $\mathbf{W_2}$; b_1^2, b_2^2, b_3^2 are the deviations from the hidden layer to the output layer, recorded as $\mathbf{B_2}$; N is the number of neurons in the hidden layer.

(4)   Optimize the particles.

Giving the weight values, deviations, and the number of hidden layer neurons to the neural network every time, the particles update and import the structure parameters into the neural network. Take the two-norm difference between the predicted data and the actual data output as the evaluation criterion for every particle. The smaller the value, the closer the particle is to the optimal solution.

During this process, the current optimal position P of a single particle, which has the smallest evaluation value for a single particle, and the current optimal position G of the entire particle swarm, which has the smallest evaluation value over the whole group are recorded, and the particle position is continuously updated with the target of optimal position. The overall process is shown in Figure 14.

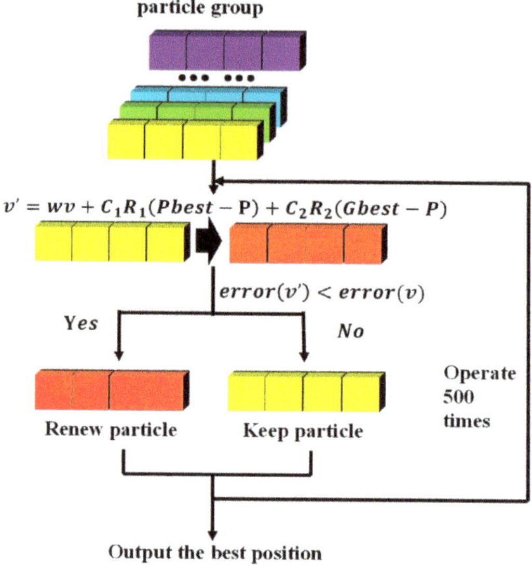

Figure 14. Process of particle optimization.

The function of velocity update is as follows:

$$v = w \times v + C_1 r_1 (Pbest - X) + C_2 r_2 (Gbest - X) \tag{17}$$

The function of position update is as follows:

$$P = P + v \tag{18}$$

where w is the inertia factor, C_1, and C_2 are individual learning factors and social learning factors, respectively, and the value range of r_1 and r_2 is (0, 1).

After iterative calculation, the particle with the lowest fitness is output, and the weight values, deviation, and a number of hidden layer neurons stored within the particle are just the optimal ones for the BP neural network.

2.   Calculate the best combination of structure parameters.

The entire process is shown in Figure 15.

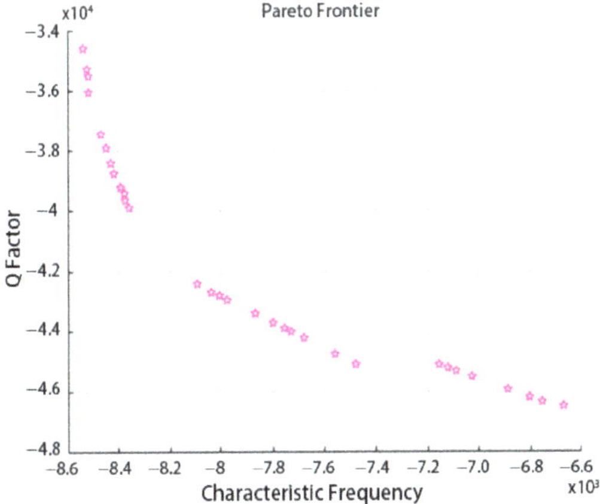

Figure 15. The pross of NSGA-II.

(1) Pick up the information of the particle the output from 1. Give the best weight values, deviation, and a number of hidden layer neurons to the BP neural network; it will be the constraint function of NSGA-II that will be used in (2).

(2) Set the BP neural network as the constraint function of NSGA-II, and set $\mathbf{AS} = \mathbf{b}$ as a constraint condition.

(3) Selecting the structure-parameter combinations that participate in genetic variation by elite strategies; Then, cod the structure-parameter combinations selected to simulate binary crossover and polynomial mutation, and put the offspring and the parent of structure-parameter combinations together. The combined structural parameter combinations were sorted via a fast non-dominant sorting algorithm, and a new parent is selected.

Repeat these 500 times, and output the Pareto frontier. The Pareto frontier is shown in Figure 16.

Figure 16. The Pareto frontier.

3. Results

We need a higher Q factor, a higher frequency difference between the frequency of ambient noise and the characteristic frequency of the operating mode, and a higher frequency difference between the characteristic frequency of other low-frequency modes and the characteristic frequency of the operating mode.

According to Figure 16, the Pareto frontier can be divided into four groups: the first is from point 1 to point 4; the second is from point 5 to point 12; the third is from point 13 to point 23; the fourth is from point 24 to point 31. Points in group 1 have the highest Q factor and the lowest frequency difference, and, among these points, the difference in the Q factor is unusually small, while the difference in frequency difference is relatively large.

The difference in the Q factor among the points of the second group is second only to the first group; however, the difference in the Q factor among the points of this group is obviously larger than it of the first group, while the differences in frequency difference are almost the same.

For the third group and the fourth group, differences in the Q factor and frequency difference among them are less distinctive. However, compared with the first group and the second group, they have a considerable variation in the Q factor and frequency difference over their points inside. Additionally, the Q factor of the points in the third group and fourth group are all far below the points in the first group and second group.

As analyzed above, point 12, the last point in the second group, is selected as the best combination of the structure parameters, which corresponds to the best performance parameters. It has a Q factor of 42,382 and a frequency difference of 8323. Additionally, its structure parameters are R = 408.726, r = 406.855, t_0 = 0.712, h = 365.505, a = 54.415.

Taking the structure parameters of the 12th point into finite element simulation, it came that the Q factor is 42,454, and the frequency difference is 8539, which is very similar to the result from PSO-BP and NSGA-II; thus, this method is proved to be available.

3.1. Local Analysis

Local analysis is operated to ensure that point 12 does not fall into a locally optimal solution.

Some of the structure parameters and the performance parameters corresponding to them around point 12 are listed in Table 4.

Table 4. The combinations of structure parameters around point 12 at the local range.

Difference in R (μm)	Difference in r (μm)	Difference in H (μm)	Difference in h (μm)	Difference in a (μm)	Characteristic Frequency	Q Factor	Frequency Difference
−0.05	0	0	0	0	31,537.99	41,554.92	6556.38
0.07	0	0	0	0	30,954.05	43,323.97	10,578.55
0	−0.09	0	0	0	30,899.5	43,650.8	11,220.80
0	0.06	0	0	0	31,857.08	41,908.96	7877.25
0	0	0	−0.08	0	30,470.41	42,853.35	8775.74
0	0	0	0.04	0	30,857.08	42,059.67	8877.25
0	0	−0.5	−0.5	0	31,246.31	42,265.57	8232.49
0	0	2.5	2.5	0	31,357.19	42,211.96	8381.35
0	0	0	0	−1.2	30,588.4	42,335.84	8910.82
0	0	0	0	2	32,744.45	42,225.87	6780.15

Additionally, the variation of performance parameters at this local range is shown in Figure 17.

It is obvious that whichever structure parameter is changed, point 12 is located in the crossover point and knee point. When a structure parameter is changed, if point 12 did not own the highest Q factor, it is inevitable that it owns the highest frequency difference. So, point 12 is the best solution to obtain the best performance parameters at the local range.

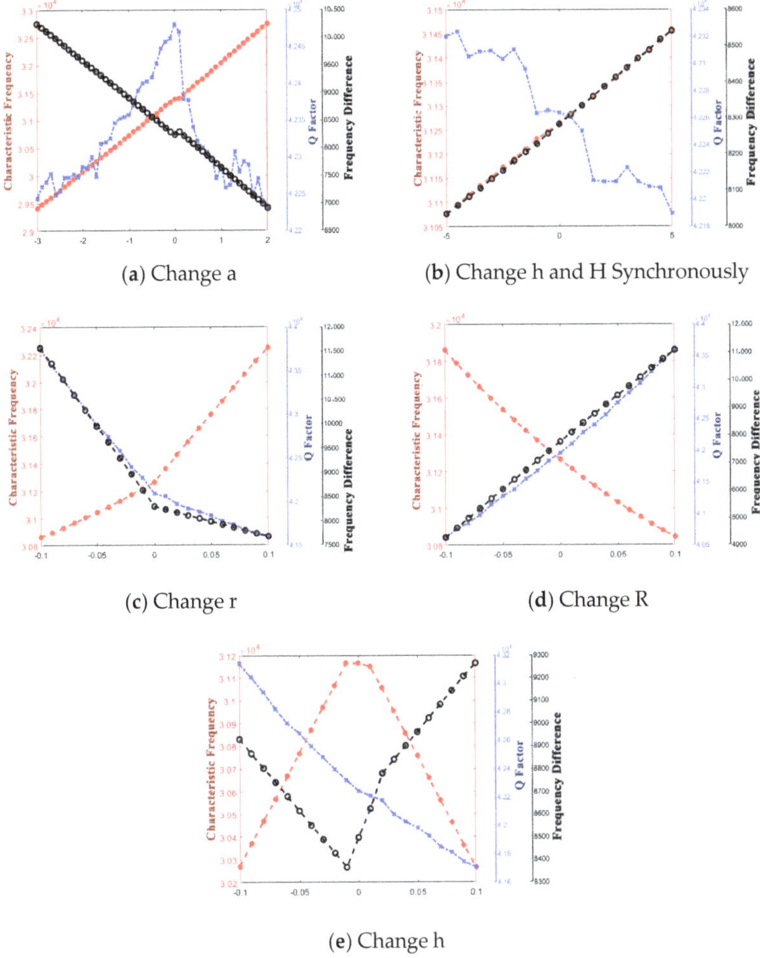

(**a**) Change a

(**b**) Change h and H Synchronously

(**c**) Change r

(**d**) Change R

(**e**) Change h

Figure 17. Variation of performance parameters at the local range around point 12.

3.2. Global Analysis

A global analysis is operated to ensure that point 12 is the best solution actually.

The structure parameters and the performance parameters corresponding to them in the selected value range are listed in Table 5.

Additionally, the variation of performance parameters at the global range is shown in Figure 18.

It is obvious that whichever structure parameter is changed, point 12 is located in the crossover point and knee point. When a structure parameter changes, if point 12 did not own the highest Q factor, it is inevitable that it has the highest frequency difference. So, point 12 is the best solution to obtain the best performance parameters at the global range. It has a Q factor of 42,454 and a frequency difference of 8539. Additionally, its structure-parameters are R = 408.726, r = 406.855, t_0 = 0.712, h = 365.505, and a = 54.415. So, when the thin film deposition method is operated, the best structure has the combination of structure parameters as mentioned above.

Table 5. The combinations of structure parameters around point 12 at the global range.

Difference in R (μm)	Difference in r (μm)	Difference in H (μm)	Difference in h (μm)	Difference in a (μm)	Characteristic Frequency	Q Factor	Frequency Difference
−0.5	0	0	0	0	28,689.24	36,126.38	8078.58
0.2	0	0	0	0	30,569.92	45,706.89	14,725.43
0	−0.4	0	0	0	30,490.75	50,160.07	20,698.64
0	0.3	0	0	0	29,136.87	38,791.72	4486.53
0	0	0	−0.35	0	27,749.8	46,513.92	10,631.14
0	0	0	0.4	0	35,183.68	41,293.17	5850.65
0	0	−20	−20	0	30,533.86	42,578.94	7373.49
0	0	32	32	0	32,501.35	41,883.58	10,083.70
0	0	0	0	−18	21,794.66	33,747.67	17,578.09
0	0	0	0	7	36,686.79	41,510.09	2971.12

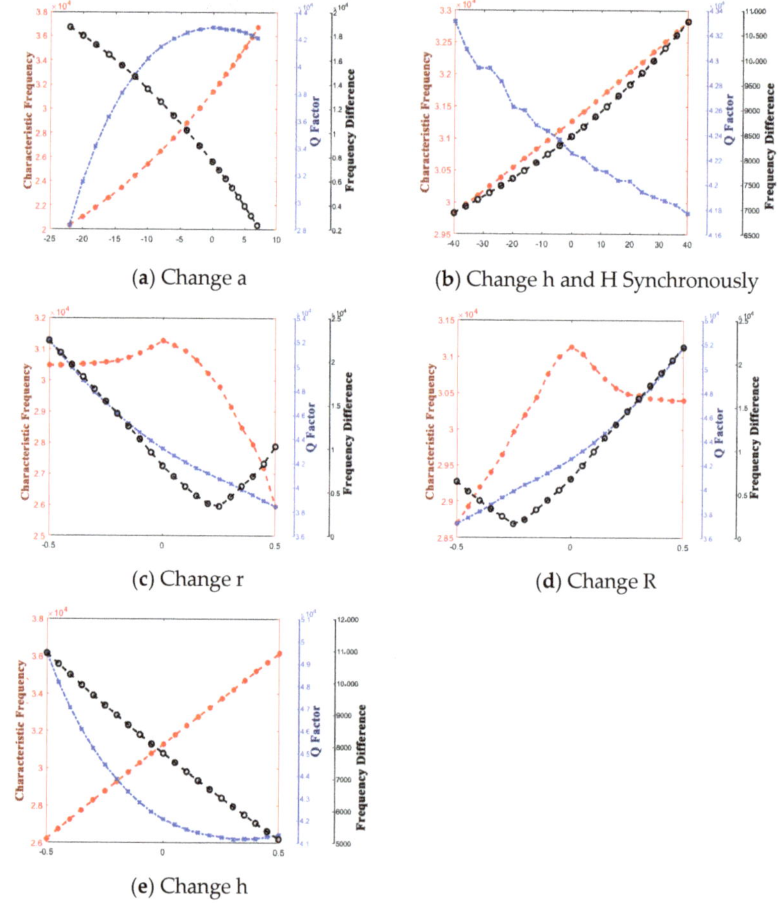

(**a**) Change a

(**b**) Change h and H Synchronously

(**c**) Change r

(**d**) Change R

(**e**) Change h

Figure 18. Variation of performance parameters at the global range around point 12.

A comparison between the simulation results of this study and experimental data from the relevant literature is as Table 6 shows.

Table 6. The comparison between the simulation results of this study and experimental data from the relevant literature.

Radius of Shell (μm)	Thickness of Shell (μm)	Depth of Shell (μm)	Radius of Anchor (μm)	Characteristic Frequency of Operating Mode (Hz)	Frequency of Spurious Mode (Hz)	Q Factor	Remark
400	1.8	—	—	28,012.21	—	14,365	[7]
650	1.7	160	—	13,649	—	22,000	[19]
408.73	1.87	365.51	54.42	31,016.25	39,555.27	42,454	This study

As Table 6 shows, the structure parameters of the polysilicon resonator given by this study have the highest Q factor among the resonators presented in reference [7,19], which is up to 42,454. The difference between the characteristic frequency of the operating mode and the frequency of the spurious mode of this resonator is 8539, ensuring that the frequency coupling between the operating mode and spurious mode is very small, which makes the influence on the hemispherical resonator generated by the spurious mode, and can be reduced to a lower level. Compared to the hemispherical resonator presented in reference [19], the radius and the thickness of the shell have a slight difference with these two parameters of the hemispherical resonator in this study. However, there is a major gap in the Q factor. Besides the differences in the material properties coming from simulation and experiment, the influence that the variety regulation mentioned above, between the structure parameters and the performance parameters, plays an important role. According to the analysis of Figure 5b, as the thickness of the shell increases, the characteristic frequency of the operating mode and the Q factor both increase continually because of the lengthening of the delivery time of heat flow, which results in the decrease in energy loss in a vibration period.

According to the analysis of Figure 9b, there is a decrease in the Q factor when the depth of the shell increases, while the characteristic frequency of the operating mode has a different trend.

There might be a disparity in the depth of the shell between reference [7] and this study. Additionally, for reference [19], according to the analysis of Figures 5 and 10, when the radius of the resonator is up to 650 μm, different from the variation trend of the characteristic frequency of operating mode, there is a decrease in Q factor when the thickness and depth of the resonator decrease, since a decrease occurs in the delivery time of heat flow, which is the cause for the increase in energy loss in a vibration period.

Since Table 6 shows that the performance parameters of the hemispherical resonator obtained via the method referring to this study are better than those of the hemispherical resonators presented in references [7,19], the feasibility that this method can be applied to the design and optimization under the limitation of a specific manufacturing process was proven.

4. Conclusions

As analyzed in Section 3, whether it is a local range or a global range, point 12 is the best solution. At the numerical intervals that we have set, the best combination of structure parameters is obtained without experimental processing. Proved by the simulation at the local range and global range, R = 408.726, r = 406.855, t_0 = 0.712, h = 365.505, and a = 54.415 is the best solution in the range of structure parameters that Table 2 actually shows.

This study provided an optimization formula for micro-resonators based on PSO-BP and NSGA-II under the limitations of specific processing conditions, avoiding experimental processing, and can be used in any one of the manufacturing processing. Obviously, it will make a difference in guiding the optimization of MEMS hemispherical resonator gyroscopes.

The structure parameters that mainly affect the performance parameters of the hemispherical shell resonator are obtained by analyzing the thermoelastic damping losses. The finite element simulations with variables controlling are used to obtain a dataset of the performance parameters of the hemispherical resonator when the structure parameters vary. By optimizing the weight, deviation, and number of hidden layer neurons via PSO, as Section 3 claims, the problem of the BP neural network that falls into a locally optimal solution is easily resolved. Finally, the limitations of molding defects under the process conditions are considered constraints, and the optimized BP neural network is used as the objective function to obtain the best structure parameters.

At the end of Section 3, validated by simulation, the result of the simulation is almost similar to the result of NSGA-II output; we prove that this approach to optimize the resonator is available. It solves the problem that it is difficult to design the best structure under a specific process because the structure parameters are under the limitation of process conditions.

However, there are also some limits in this study. First, the weight of the Q factor and frequency difference is not clear, although the Pareto frontier gives the non-dominated optimal solution set, to what an extent that Q factor and frequency difference occupies the comprehensive performance should be quantized clearly; another limit is that what we study is in a selected numerical interval, in different numerical intervals, there might be a great difference among the mappings, so, the numerical intervals selected in an optimization should not be too large.

Moreover, there would be a sudden change when the hemispherical resonator changes from a standard one to a nonstandard one because that the sudden change in the structure results in a sudden change in the path length for the irreversible heat flow.

For the next work, the characteristic frequency of the operating mode can be limited under 10 kHz or 15 kHz as it is a little high for the frequency of 31 kHz from this study; another job is to discover the particular cause of the sudden change when the hemispherical resonator changes from a standard one to a nonstandard one.

Author Contributions: Conceptualization, J.L. (Jinghao Liu), X.Z. and Y.S.; methodology, J.L. (Jinghao Liu) and X.Z.; software, J.L. (Jinghao Liu) and Y.S.; validation, X.Z.; formal analysis, Y.S.; investigation, P.L.; resources, P.L.; data curation, Q.Q., J.L. (Jialuo Liao), Z.G. and M.L.; writing—original draft preparation, Q.Q.; writing—review and editing, J.L. (Jinghao Liu); visualization, M.L.; supervision, P.L.; project administration, P.L.; funding acquisition, X.Z. and Y.S. All authors have read and agreed to the published version of the manuscript.

Funding: This research was funded by the Taishan Scholars Program of Shandong Province No. tsqn 201909108, the Key Research and Development Project of Zibo City under Grant No. 2020SNPT0088, and the Natural Science Foundation of Shandong Province, China, Grant No. ZR2021MF042.

Data Availability Statement: Not applicable.

Acknowledgments: The authors deeply thank the support from the Taishan Scholars Program of Shandong Province of China (201909108), the Key Research and Development Project of Zibo City under Grant of Zibo City (2020SNPT0088), and the Natural Science Foundation of Shandong Province of China (ZR2021MF042) and Shandong Provincial Key Laboratory of Precision Manufacturing and Non-traditional Machining.

Conflicts of Interest: The authors declare no conflict of interest.

References

1. Ranji, A.R.; Damodaran, V.; Li, K.; Chen, Z.; Alirezaee, S.; Ahamed, M.J. Recent Advances in MEMS-Based 3D Hemispherical Resonator Gyroscope (HRG)—A Sensor of Choice. *Micromachines* **2022**, *13*, 1676. [CrossRef] [PubMed]
2. Wei, L.; Kuai, X.; Bao, Y.; Wei, J.; Yang, L.; Song, P.; Zhang, M.; Yang, F.; Wang, X. The Recent Progress of MEMS/NEMS Resonators. *Micromachines* **2021**, *12*, 724. [CrossRef] [PubMed]
3. Zhuang, X.; Wang, S.; Li, P. Development Status and Trend Analysis of Microscale Hemispherical Gyroscope Fabrication Technologies. *Navig. Control* **2019**, *18*, 52–60.
4. Eklund, E.J.; Shkel, A.M. Glass Blowing on a Wafer Level. *J. Microelectromech. Syst.* **2007**, *16*, 232–239. [CrossRef]
5. Cho, J.Y.; Najafi, K. A high-Q all-fused silica solid-stem wineglass hemispherical resonator formed using micro blow torching and welding. In Proceedings of the IEEE International Conference on Micro Electro Mechanical Systems, Estoril, Portugal, 18–22 January 2016.

6.   Xing, Y. Process and Experimental Study on Micro Fused Silica Hemispherical Resonator. Master's Thesis, Shanghai University, Shanghai, China, 2017.

7.   Zhuang, X.; Chen, B.; Li, P.; Wang, X.; Cao, W.; Ding, J. Fabrication of microscale axial symmetric three-dimensional curved shell resonators based on silicon isotropic wet etching. *Opt. Precis. Eng.* **2019**, *27*, 1293–1300. [CrossRef]

8.   Gray, J.M.; Houlton, J.P.; Gertsch, J.C. Hemispherical micro-resonators from atomic layer deposition. *J. Micromech. Microeng.* **2014**, *2014*, 125028. [CrossRef]

9.   Sorenson, L.D.; Gao, X.; Ayazi, F. 3-D micromachined hemispherical shell resonators with integrated capacitive transducers. In Proceedings of the 2012 IEEE 25th International Conference on Micro Electro Mechanical Systems (MEMS), Paris, France, 29 January–2 February 2013.

10.   Lu, K. Research on the Structure Design and Processing Technology of Wineglass Micro Shell Vibrating Gyroscope. Ph.D. Thesis, National University of Defense Technology, Changsha, China, 2015.

11.   Lin, J.; Xu, Y.; Song, J.; Tang, J.; Fang, Z.; Wang, M.; Fang, W.; Cheng, Y. Femtosecond laser direct writing of high-Q micro resonators in glass and crystals. In *Laser Resonators, Microresonators, and Beam Control XVII*; SPIE: San Francisco, CA, USA, 2015.

12.   Li, W.; Xiao, D.; He, H.; Hou, Z.; Wu, X.Z. Single-crystal diamond micro-cupped resonators made by laser ablation. *Microsyst. Technol.* **2016**, *22*, 2603–2610. [CrossRef]

13.   Zhang, T.; Deng, J.; Yang, F.; Mei, S.; Xiao, K.; Lin, B.; Han, S.; Lin, C. Fabrication of Micro-Hemispherical Gyroscope Based on Fractional Blowing Technology. *Piezoelectrics Acoustooptics* **2021**, *43*, 146–148.

14.   Boyd, C.; Woo, J.K.; Cho, J.Y.; Nagourney, T.; Darvishian, A.; Shiari, B.; Najafi, K. Effect of drive-axis displacement on MEMS birdbath resonator gyroscope performance. In Proceedings of the IEEE International Symposium on Inertial Sensors and Systems, Laguna Beach, CA, USA, 27–30 March 2017.

15.   Gao, Y.; Ding, X.; Zhang, H. High-Temperature Blowing Process Simulation and Study on Shape Influence Mechanism of Micro-Hemispheric Shell Harmonic Oscillator. *J. Test Meas. Technol.* **2023**, *37*, 5–10+17.

16.   Shi, Y. Fabrication Technology and Quality Factor Improvement for Fused Silica Micro Hemispherical Resonator. Ph.D. Thesis, National University of Defense Technology, Changsha, China, 2021.

17.   Xiaowen, L. Frequency Splitting Analysis of Hemispherical Harmonic Oscillator and Study on Vibration Energy Loss. Ph.D. Thesis, Harbin Institute of Technology, Harbin, China, 2020.

18.   Zhuang, X.; Bing, J.; Cao, W. Nondestructive Detection Technology of Silicon Hemispherical Deep Cavity Structure. *Navig. Control* **2019**, *18*, 89–95.

19.   Zhuang, X.; Chen, B.; Wang, X.; Yu, L.; Wang, F.; Guo, S. Microsacle PolySilicon Hemispherical Shell Resonating Gyroscopes with Integrated Three-dimensional Curved Electrodes. In Proceedings of the 19th Annual Conference and 8th International Conference of Chinese Society of Micro/Nano Technology (CSMNT), Dalian, China, 26–29 October 2017.

 *micromachines*

Article

Microwave Dual-Crack Sensor with a High Q-Factor Using the TE$_{20}$ Resonance of a Complementary Split-Ring Resonator on a Substrate-Integrated Waveguide

Yelim Kim [1], Eiyong Park [1], Ahmed Salim [2], Junghyeon Kim [1] and Sungjoon Lim [1,*]

[1]   School of Electrical and Electronics Engineering, Chung-Ang University, Seoul 06974, Republic of Korea
[2]   School of Engineering, University of British Columbia, Kelowna, BC V1V 1V7, Canada
*   Correspondence: sungjoon@cau.ac.kr

Abstract: Microwave sensors have attracted interest as non-destructive metal crack detection (MCD) devices due to their low cost, simple fabrication, potential miniaturization, noncontact nature, and capability for remote detection. However, the development of multi-crack sensors of a suitable size and quality factor (Q-factor) remains a challenge. In the present study, we propose a multi-MCD sensor that combines a higher-mode substrate-integrated waveguide (SIW) and complementary split-ring resonators (CSRRs). In order to increase the Q-factor, the device is miniaturized; the MCD is facilitated; and two independent CSRRs are loaded onto the SIW, where the electromagnetic field is concentrated. The concentrated electromagnetic field of the SIW improves the Q-factor of the CSRRs, and each CSRR creates its own resonance and produces a miniaturizing effect by activating the sensor below the cut-off frequency of the SIW. The proposed multi-MCD sensor is numerically and experimentally demonstrated for cracks with different widths and depths. The fabricated sensor with a TE$_{20}$-mode SIW and CSRRs is able to efficiently detect two sub-millimeter metal cracks simultaneously with a high Q-factor of 281.

Keywords: complementary split ring resonator; metamaterial-based sensor; quality factor; multi-crack detection; higher-mode substrate-integrated waveguide

Citation: Kim, Y.; Park, E.; Salim, A.; Kim, J.; Lim, S. Microwave Dual-Crack Sensor with a High Q-Factor Using the TE$_{20}$ Resonance of a Complementary Split-Ring Resonator on a Substrate-Integrated Waveguide. *Micromachines* **2023**, *14*, 578. https://doi.org/10.3390/mi14030578

Academic Editors: Yong Ruan, Weidong Wang and Zai-Fa Zhou

Received: 19 January 2023
Revised: 24 February 2023
Accepted: 27 February 2023
Published: 28 February 2023

1. Introduction

Surface cracks in the metal caused by fatigue and corrosion, which can lead to significant problems in both small metallic objects and large structures, are difficult to detect and repair. Recently, various metal crack detection (MCD) techniques have been developed, including destructive tests such as acid corrosion, sulfur printing, and pinhole tests. However, non-destructive testing (NDT) has received significant attention because destructive tests can cause undesired damage and are limited in detecting cracks under coatings, such as rust, paint, corrosion protection, and composite laminate [1–3]. NDT techniques include advanced ultrasound, acoustic emissions, eddy currents, and optical fiber techniques, all of which offer high-resolution crack detection. However, they have disadvantages such as high complexity, high cost, large-sized equipment, and manufacturing difficulties [4–12]. In contrast, microwave-based NDT employing microstrips, waveguides, meta-resonators, radio-frequency (RF) identification, and antennas are low-cost, have a smaller sensor size, and allow for easier fabrication [13–20]. Despite these advantages, microwave-based NDT is limited in its ability to detect a variety of crack shapes and multiple cracks simultaneously and generally requires an increase in sensitivity.

The quality factor (Q-factor) is a key parameter for sensitivity; thus, various RF sensor studies have sought to develop sensor systems with a high Q-factor [21–23]. High Q-factor RF sensors can be realized using active or passive feedback systems. Assisting an active feedback loop using an active microwave device can dynamically increase the Q-factor; however, DC adjustment and calibration are important issues for this setup that need to be solved. Another problem associated with active microwave sensors is their limited operating temperature range [24–29]. On the other hand, passive structures with high Q-factor are generally implemented using reduced capacitance or increased inductance, complicating the structure. As a result, it is difficult to simultaneously detect multiple cracks and produce a high Q-factor because multi-crack detection sensors that requires independently operating resonances have unique structures such as multiband filters or multiband antennas [30–34]. Sensors that can detect two materials simultaneously and independently have gained significant research attention because they can significantly increase the sensing accuracy and reduce the scanning time for large areas while reducing power consumption [35–46]. In this paper, we propose a novel method that can increase the number of cracks that can be detected and produce a high Q-factor by integrating complementary split-ring resonators (CSRRs) with a higher-mode substrate-integrated waveguide (SIW).

A CSRR is a metamaterial structure that has an etched SRR on the opposite side of the signal line. Metamaterials (MM) are materials with the negative property of permeability and were first shown experimentally in 1999 as split ring resonators (SRR). The SRR and CSRR have been used for various RF applications such as antennas, sensors, filters, and transmission lines [46–48]. In CSRR, we can control the inductance and capacitance by the slots and line of CSRR. For example, when a voltage difference is excited in the CSRR, capacitance is induced in the slot by the electric field (E-field), and inductance is induced in the metal strip and conducting island by the magnetic field (H-field). For example, when a voltage difference is excited in the CSRR, capacitance is induced in the slot by the electric field (E-field), and inductance is induced in the metal strip and conducting island by the magnetic field (H-field) [49–51]. CSRRs can be used for multiple sensing because, when independently patterned CSRRs are employed, each pattern generates its own resonance frequency. In [42], sixteen CSRRs were loaded on four signal lines for multi-sensing, and in [45], coupled CSRRs were used for a dual- and triple-notch filter, and three CSRRs with a cavity were used in [52]. Additionally, the external field and Q-factor (~100) of CSRRs are helpful for MCD. Despite the already high Q-factor for CSRRs, various studies proposed new CSRR structures to increase the frequency-shifting sensitivity or the Q-factor further [52–63]. In [56,59,64], increasing the capacitance or inductance of the CSRR was used to increase the frequency-tuning sensitivity, and a higher Q-factor was realized by increasing the number of CSRRs [54,62,63] and using a T-shaped resonator [64–67].

In this paper, we produce a high Q-factor (~200) for a multi-crack sensor by integrating CSRRs with a higher-mode SIW. A SIW is a planar structure that offers additional functionalities over traditional waveguides, and attempts have been made to combine meta-material structures with a SIW to increase the Q-factor and sensitivity or reduce the sensor size [68–74]. However, combining CSRRs with a higher-mode SIW to separate and concentrate E-fields is proposed for the first time in the present study to enable multi-crack detection with a high Q-factor. When the independent CSRRs are loaded onto the SIW, each CSRR develops a stronger E-field distribution exterior to the SIW plane at each resonance frequency. To demonstrate our approach, we fabricated a dual-crack sensor using two CSRRs on a TE_{20}-mode SIW. We loaded one CSRR at each hotspot to enhance the individual E-fields and simultaneously detected two metal cracks. The observed results for the proposed crack sensor confirmed that the sensor could detect different crack widths and depths simultaneously and independently.

2. Proposed Sensor Design

Figure 1 summarizes the principles underlying the proposed multi-crack detection sensor with a high Q-factor. First, we used a higher-mode SIW to form a separated E-field distribution, and the SIW subsequently enhanced the Q-factor of the CSRRs (Figure 1a). Next, independent CSRRs were positioned in the separated and concentrated E-fields on the TE_{20}-mode SIW. Each CSRR produces an independent resonance, and the resonance frequency strengthens the E-field of the SIW (Figure 1b,c). As a result, changes in the metal surface, such as cracks or slots, sensitively affect the reflective coefficient due to the stronger external field for each CSRR (Figure 1d).

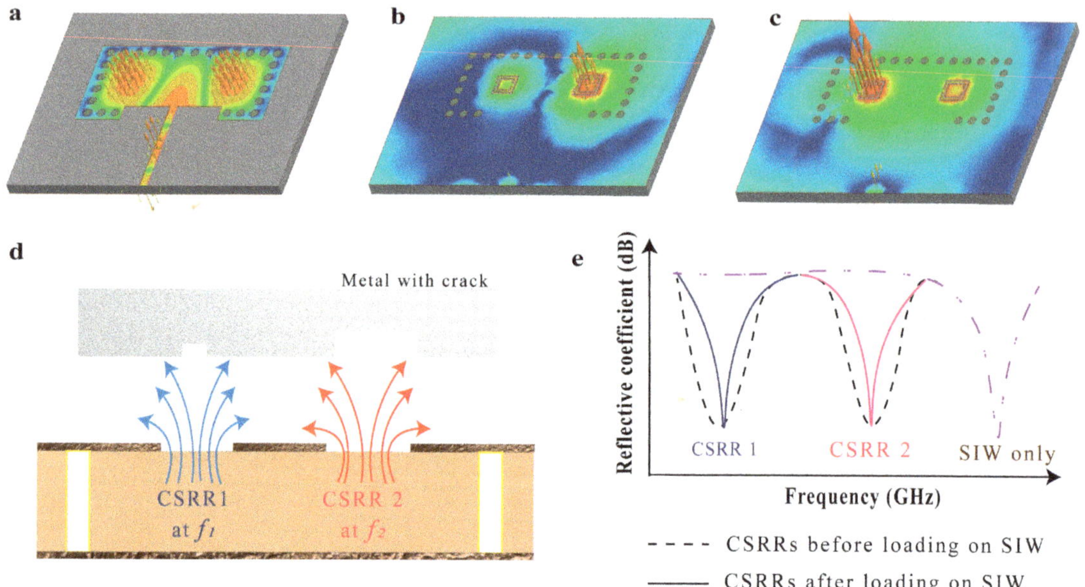

Figure 1. Proposed multi-crack detecting sensor with a high Q-factor: (**a**) TE_{20} mode SIW only E-field vector distribution. E-field vector distribution after the integration of the CSRRs (**b**) at 4.69 GHz and (**c**) 5.49 GHz. The E-field intensity range of (**a–c**) is the same. Concept illustration of (**d**) independent multi-crack detection and (**e**) the reflective coefficients before and after CSRR loading.

For example, when a crack is near CSRR1, the field near this CSRR changes, affecting the reflective coefficient for its resonance frequency. On the other hand, when a crack is near CSRR2, the crack only affects the reflective coefficient for the resonance frequency of CSRR2. When cracks are near both CSRRs, the reflective coefficients both change. In particular, loading the CSRR onto a higher-mode SIW has some advantages. Firstly, a higher-mode SIW enhances the Q-factor of the CSRR by strengthening the electromagnetic field; thus, the proposed sensor has a sub-millimeter sensing resolution and can distinguish between various crack shapes. Secondly, the use of CSRRs reduces the physical size of the SIW while maintaining its advantages because the higher capacitance and inductance of the CSRR decrease the operating frequency of the SIW (Figure 1e).

Table 1 compares the performance of the proposed crack sensor with that of conventional multi-detection RF sensors and crack detection sensors using CSRRs. These approaches have received significant attention because multi-detection capability significantly reduces the scanning time over wide areas and reduces power consumption. In [44], a power divider was used to produce independent frequencies and increase the Q-factor, leading to the fabrication of a sensor with high accuracy for liquid material permittivity measurements. In [16], a substrate highly sensitive to temperature was used to combine temperature and crack sensing. In [18], metal crack and strain sensing were achieved using a dual-mode passive and active antenna.

For MCD, many researchers have focused on increasing the Q-factor or the number of cracks that can be detected, for which antennas and CSRRs are widely used. We previously demonstrated an increase in the Q-factor by integrating a CSRR with the dominant mode of a SIW [35]. In [36], four SRRs were used to increase the detectable number of cracks, exhibiting high sensitivity at 38 GHz. However, SRRs create E-fields on the surface, and some exhibit unstable detection performance with a sensing resolution that varies from 0.1 mm to 0.5 mm, depending on the crack position. In [14,41], higher-mode antennas were introduced to produce independent frequencies and detect the tilting angle of metal cracks. In [44], a spoof surface plasmon polariton sensor was used for multiple crack detection by incorporating liquid channels. The sensor had high accuracy, sensing resolution, and sensitivity, but liquid switches have a slow switching speed and require an additional system for movement and storage. Although the sensor in [13] had the capability to detect 12 cracks simultaneously and independently, it had a low Q-factor and resolution. In contrast, the highly sensitive sensor proposed in the present study has a high Q-factor and sensing resolution in terms of both width and depth. In particular, the sensor can detect crack depths at a resolution of 0.2 mm due to the stronger external E-field.

Table 1. Summary of the proposed and previously reported multi-crack detection sensors.

[Ref] Year	No. of Cracks Detected	Sensing Target	Sensing Techniques	Crack Sensing Resolution (mm)		Sensitivity (MHz/mm)		Measured Q-Factor
				Width	Depth	Width	Depth	
[39] 2020	2	Liquid material permittivity	Power divider + SRRs	N/A	N/A	N/A	N/A	280
[16] 2021	2	Temperature and metal cracks	Antenna + temperature-sensitive substrate	0.10	N/A	16.6	N/A	17
[18] 2020	2	Metal cracks and strain	Passive and active mode antenna	0.25	N/A	50	N/A	N/A
[35] 2012	1	Metal cracks	CSRR	0.10	N/A	200	N/A	25
[50] 2017	1	Metal cracks	SIW + CSRR	0.20	500	50	1	224*
[36] 2021	4	Metal cracks	Four SRRs	0.10–0.50	N/A	1300	N/A	N/A
[40] 2018	2	Metal cracks	Higher-mode patch antenna	1.00	N/A	140	N/A	N/A
[14] 2019	2	Metal cracks	Higher-mode patch antenna	0.20	N/A	45	N/A	N/A
[44] 2022	4	Metal cracks	Spoof surface plasmon polariton sensor + liquid switch	0.10	N/A	200	N/A	N/A
[42] 2021	12	Metal cracks	CSRRs	0.40	0.2	N/A (magnitude)		17
This work	2	Metal cracks	CSRRs + higher-mode SIW	0.10	0.2	200 *	130 *	281 *

* Highest value when the minimum crack is loaded.

2.1. Design of the Proposed Dual-Crack Detection Sensor

Figure 2 presents the proposed dual-crack sensor design. It consists of a three-layered bottom conductive element with a signal line, a top conductive layer with two etched CSRRs on the ground plane, and a substrate with a via that connects the top and bottom plane to produce SIW cavities. We used a 1.27 mm thick Rogers Duroid 6010 substrate with a dielectric constant (ε_r) of 10.2 and a loss tangent (tan δ) of 0.0023. Additionally, the final sensor design included an uncracked metal sheet (Figure 1c). In this paper, aluminum with a thickness of 5 mm was used by a metal sheet.

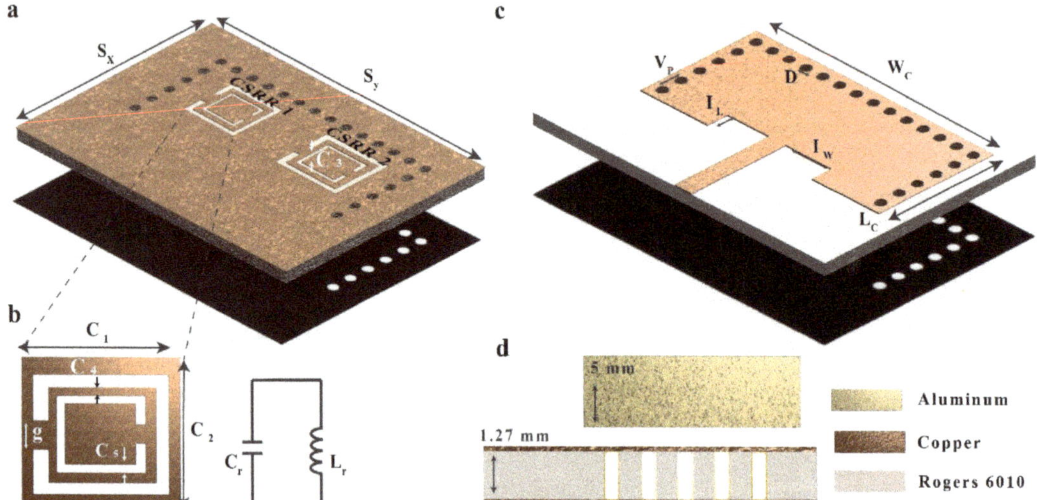

Figure 2. Proposed dual-crack detection sensor: (**a**) top view, (**b**) CSRR structure and equivalent circuit, (**c**) bottom view, and (**d**) side view.

The geometrical parameters for the SIW and CSRRs were determined using Equations (1)–(8). First, we designed the TE_{20}-mode SIW structure using Equations (1)–(3). The resonance frequency of the TE_{20}-mode SIW was calculated as

$$f_{mn0} = \frac{1}{2\sqrt{\mu\varepsilon}}\sqrt{\left(\frac{m}{W_{eff}}\right)^2 + \left(\frac{n}{L_{eff}}\right)^2},\tag{1}$$

where ε, μ, m, n, W_{eff}, and L_{eff} are the substrate permittivity, permeability, mode m index, mode n index, and effective width and length, respectively. The effective width and length were calculated as

$$W_{eff} = W_c - 1.08\frac{D^2}{V_p} + 0.1\frac{D^2}{W_C}\tag{2}$$

$$L_{eff} = L_c - 1.08\frac{D^2}{V_p} + 0.1\frac{D^2}{L_c}\tag{3}$$

where W_c and L_c are the width and length of the SIW, respectively, from the center of the vias, and D is the diameter of the vias [47,74]. Based on the SIW, the TE_{20}-mode resonant frequency for the SIW loaded with CSRRs can be expressed as

$$f_{CSRR} = \frac{1}{2\pi\sqrt{L_rC_r}}\tag{4}$$

where L_r and C_r are the inductance and capacitance of the CSRR as shown in Figure 2b [74,75], which are determined by Equations (5) and (6), respectively:

$$L_r = \frac{4.86\mu_0}{2}(c_1 - c_4 - c_5)\left[\ln\left(\frac{0.98}{\rho} + 1.84\rho\right)\right]\tag{5}$$

$$C_r = (c_1 - 1.5(c_4 + c_5))C_{pul}\tag{6}$$

where $\rho = \frac{c_4 + c_5}{c_1 - c_4 - c_5}$ and C_{pul} is the capacitance per unit length between the rings [76–80].

In addition, the proposed sensor employs CSRRs to reduce the evanescent wave effects and obtain a higher Q-factor compared with SIW-only approaches. The Q-factor at each resonant frequency can be expressed as

$$Q = R\sqrt{\frac{C_r}{L_r}} \tag{7}$$

where C_r and L_r are defined in Equations (5) and (6), respectively. A higher Q-factor increases the sensitivity, i.e., sharper resonance E-field concentrations and perturbations [77]. A large D and V_p can occur due to E-field leakage in the SIW, thus we limit these to reduce radiation losses as follows:

$$D < \frac{\lambda_g}{5}, \; V_p \leq 2D, \tag{8}$$

where λ_g is the guided wavelength.

Finally, the SIW cavities within the substrate have a width W of 25 mm, a length L of 11.1 mm, a via diameter D of 1 mm, and a center-to-center separation V_p of 1.8 mm (Figure 1d). The two CSRRs etched on the bottom of the sensor have a gap g of 0.5 mm and an external and internal ring width and ring spacing (c_3, c_4, and c_5, respectively) of 0.2 mm each. The external square width and length (c_1 and c_2) are both 2.7 mm for CSRR1 (left side, Figure 2a) and both 3.1 mm for CSRR2 (right side, Figure 2a), with the other parameters the same for the two CSRRs. The cavities are connected to a 50-Ω microstrip line with a width of 1 mm and a length of 10 mm that links to the subminiature version A (SMA) connector. The cavity inserts were designed to match the impedance with the aluminum sheet used as the CSRR loading plane, with a width I_W of 6 mm and a length I_L of 2 mm. The total substrate size is SX = 36 mm and SY = 26 mm.

2.2. Sensing Method

Figure 3 presents simulated S11 results for the proposed crack sensor. The resonant frequency of the sensor changes from 4.626 and 5.436 GHz to 4.053 and 5.208 GHz after loading the Al sheet because the addition of material that has a different dielectric constant affects the resonant frequency (Figure 3a). In this paper, we optimized the sensor via the loading material to increase its sensitivity for surface cracks. Figure 3b−d show that the proposed sensor can detect the position of multiple cracks with a width, depth, and length of $0.5 \times 0.5 \times 6.1$ mm^3. For example, when cracks are above both CSRRs, the resonance frequencies change from 4.053 and 5.208 GHz to 4.020 and 5.145 GHz for CSRR1 and CSRR2, respectively. On the other hand, when a crack is only above CSRR1, only the resonance frequency for CSRR1 changes, while the same is true for CSRR2. This is because the E-field near the CSRR is disturbed by the material, affecting the resonant frequency of the CSRR.

In addition, when a crack is present in the CSRR region, the capacitance is affected. The concentrated E-field near the CSRR flows along the crack, which operates as a capacitor. The shorter side of the crack is considered the crack width, and it can be considered the distance between capacitors. The crack depth can be considered the area of the capacitor. When the crack width increases, the capacitance decreases, and the resonant frequency increases. In addition, as the crack depth increases, the capacitance rises, and the resonant frequency falls. The resonant frequency of a CSRR in the presence of a crack is represented by Equation (9):

$$f_{CSRR} = \frac{1}{2\pi\sqrt{L_r(C_r + (C_{crack})}}, \tag{9}$$

where C_{crack} is the capacitance included by the crack. Thus, the proposed sensor detects a crack using the change in the resonant frequency. The proposed multi-crack sensor has a low fabrication cost and small waveguide cavities, and the independent sensing area can be extended because multi-E-fields in TEmn mode do not affect other sensing results [81–83].

Figures 4–7 are numerical analyses to check the sensing capability. Figure 4a–e show the sensing resolution for different crack width when the crack is positioned in the center of each CSRR. The proposed sensor offers multi-crack sensing at a high sensing resolution based on the stronger Q-factor. As a result, the sensor has a detection resolution of 0.1 mm for width and 0.2 mm for depth in both positions.

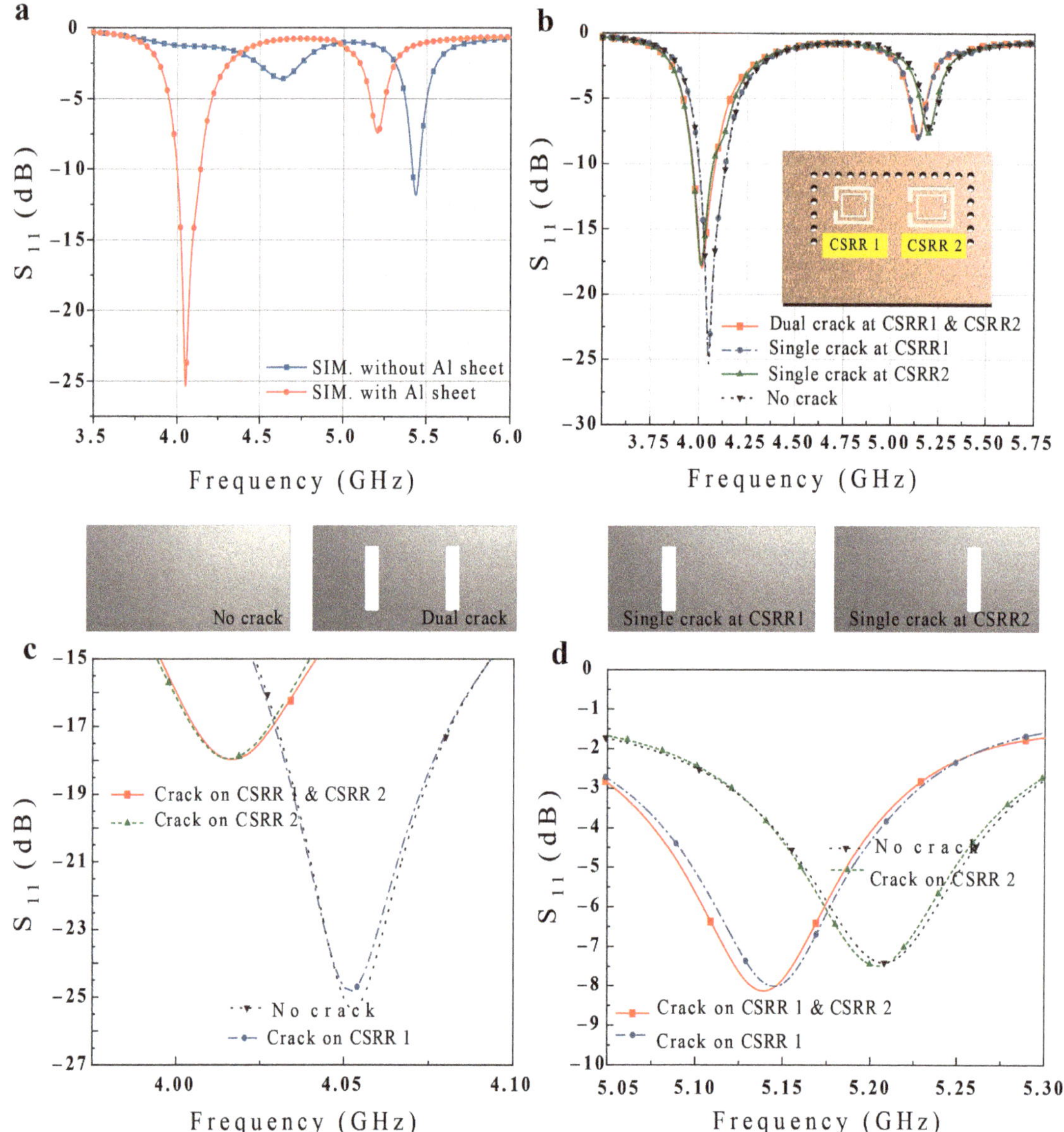

Figure 3. Simulation results for the proposed crack sensor: (**a**) S11 without an Al sheet and with an uncracked Al sheet, (**b**) S11 for different crack numbers and positions (two cracks, single crack at CSRR1, single crack at CSRR2, no crack), S11 results (**c**) in the CSRR2 resonance frequency band, and (**d**) in the CSRR1 resonance frequency band (crack width, depth, and length of $0.5 \times 0.5 \times 6.1$ mm).

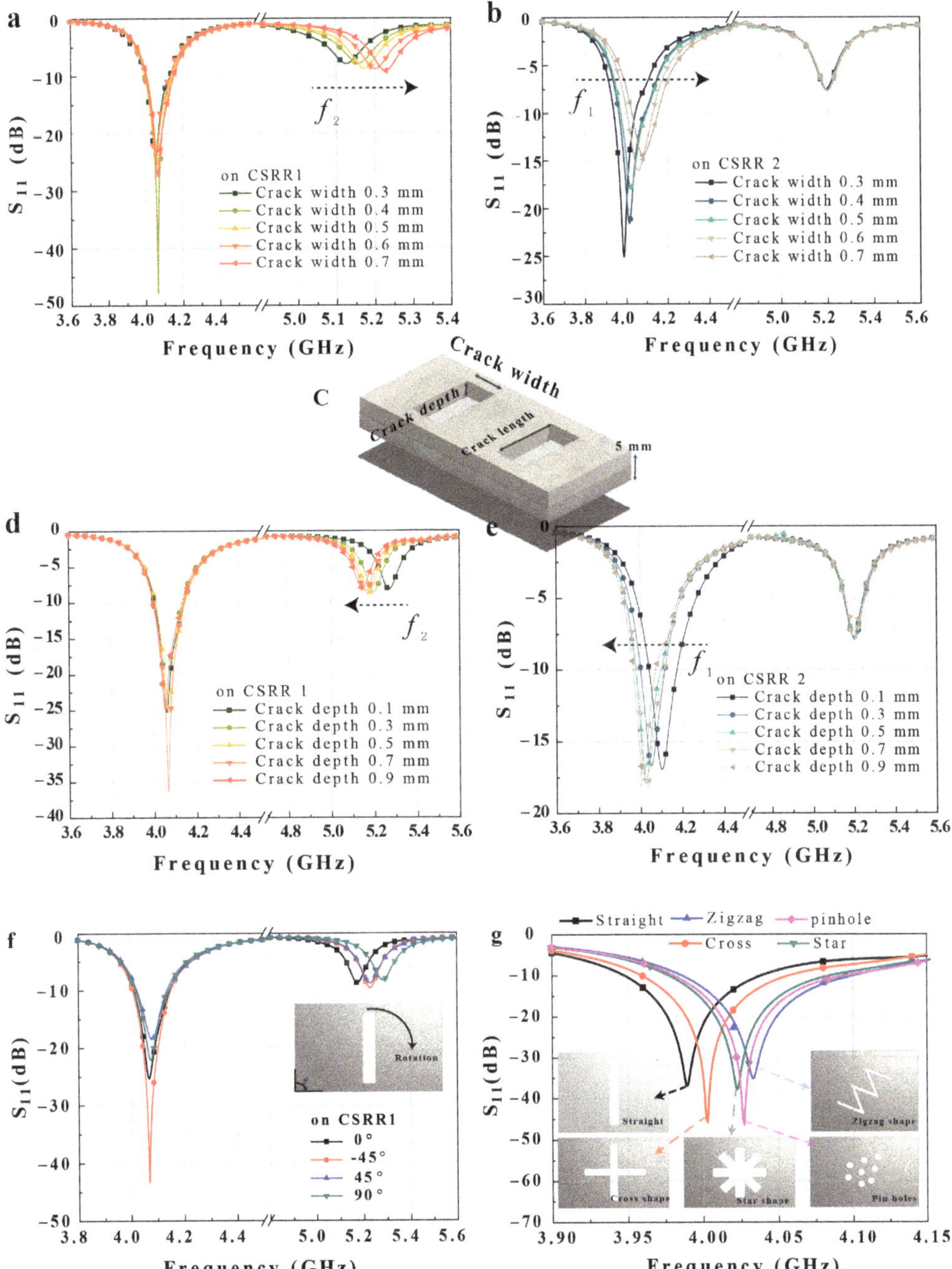

Figure 4. Simulated S11 results for the proposed dual-crack sensor for different crack widths above (**a**) CSRR1 and (**b**) CSRR2. (**c**) Fundamental metal sheet structure with two cracks with a width, depth, and length of $0.5 \times 0.9 \times 6.1$ mm. S11 results for different crack depths above (**d**) CSRR1 and (**e**) CSRR2. (**f**) S11 results for the rotation of a straight crack above CSRR1. (**g**) S11 results for five crack shapes above CSRR2.

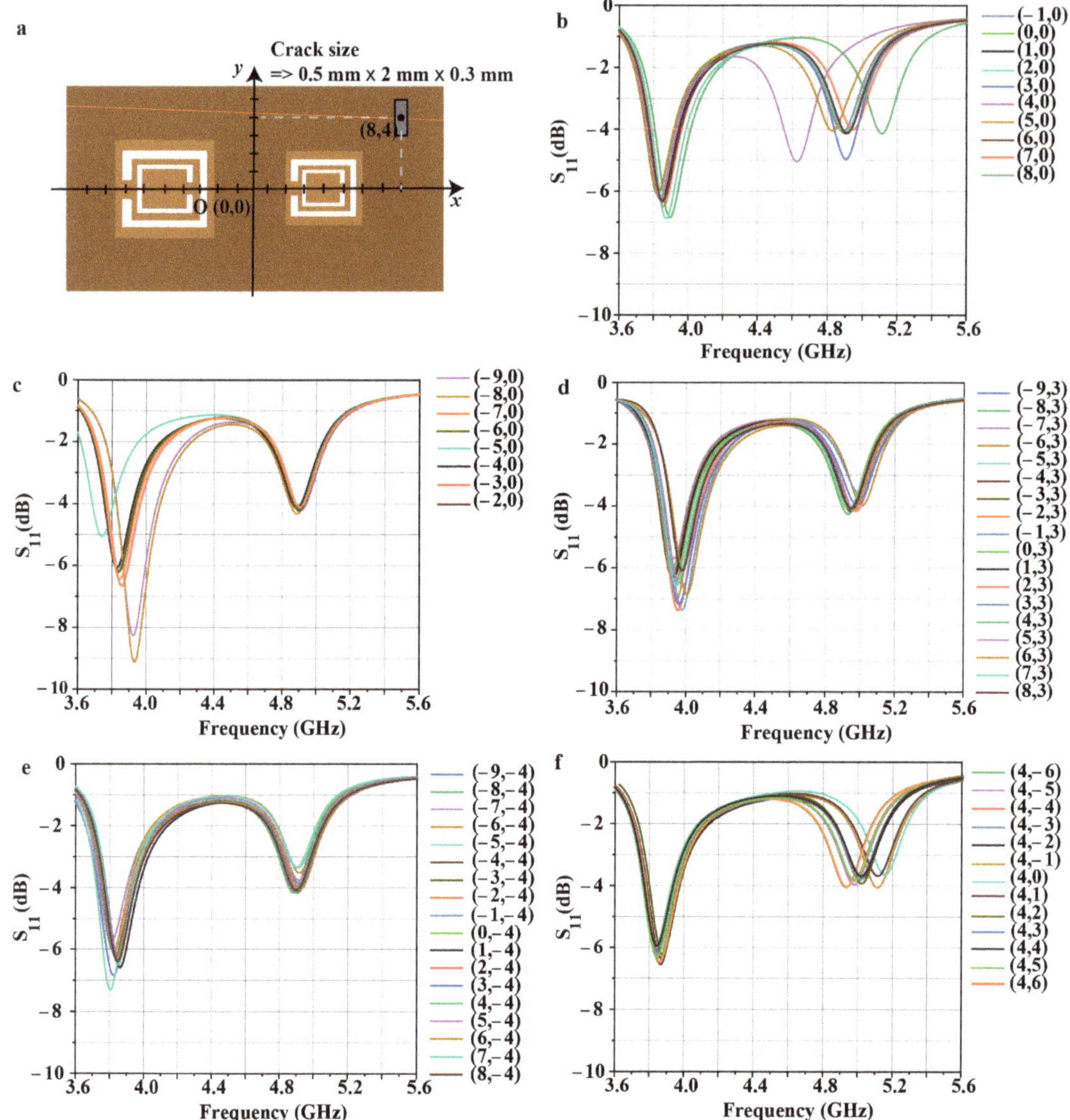

Figure 5. Simulation results for different crack locations (x, y) on the Al sheet: (**a**) coordinate of the crack; (**b**–**e**) S11 when the crack moves x-axis; (**b**) from $(-9, 0)$ to $(-2, 0)$, (**c**) from $(-1, 0)$ to $(8, 0)$, (**d**) from $(-9, 3)$ to $(8, 3)$, and (**e**) from $(-9, 4)$ to $(8, 4)$; (**f**) S11 when the crack move y-axis from $(4, -6)$ to $(4, 6)$.

Figure 4a,b display the simulated frequency responses for various straight crack widths, with a consistent length of 6.1 mm for all cracks. Figure 4a presents the simulated S11 results as the width of a single straight crack above CSRR1 increases from 0.3 to 0.7 mm at intervals of 0.1 mm with a fixed depth of 0.5 mm. The resonant frequency for CSRR1 increases from 5.126 to 5.229 GHz as the crack width increases, and the average change in frequency is 257.5 MHz/mm (Figure 4a). On the other hand, when the width of a single straight crack above CSRR2 increases from 0.3 to 0.7 mm at intervals of 0.1 mm with a fixed depth of 0.5 mm, the resonant frequency for CSRR2 increases from 3.988 to 4.083 GHz and the average frequency shift is 237.5 MHz/mm. Figure 4d,e present the simulated S11 results when the depth of a single straight crack above CSRR1 and CSRR2 increases from 0.1 to 0.9 mm at intervals of 0.2 mm. The resonant frequency of CSRR1 decreases from 5.269 to 5.145 GHz, and the average change in frequency is 142.5 MHz/mm when the crack depth is changed above CSRR1. In contrast, when the crack depths above CSRR2 increase, the resonant frequency of CSRR2 decreases from 4.103 to 4.011 GHz, and the average change in frequency is 132.5 MHz/mm. The simulation results thus demonstrate the sensitivity of the proposed sensor for the detection of straight cracks, with the ability to distinguish differences of 0.1 and 0.2 mm in width and depth, respectively.

The proposed dual sensor can detect various shapes at each CSRR due to its high sensitivity. Figure 4f,g present the simulated S11 results for five crack shapes commonly found on metal surfaces: tilted, cross, zigzag, star, and pinhole. Figure 4f presents the S11 results for different tilting angles above CSRR1. A ±45° rotation of the crack increases the resonant frequency to 5.228 GHz compared with a resonant frequency of 5.174 GHz at 0°. Similarly, a 90° horizontally oriented straight crack has a resonant frequency of 5.283 GHz. Figure 4g displays the S11 results for the straight, cross, zigzag, star, and pinhole cracks above CSRR2. Because the difference in the crack distribution affects Ccrack in Equation (9), and small differences can have a significant effect due to the concentrated E-field, the different crack patterns lead to differences in the change in the resonant frequency.

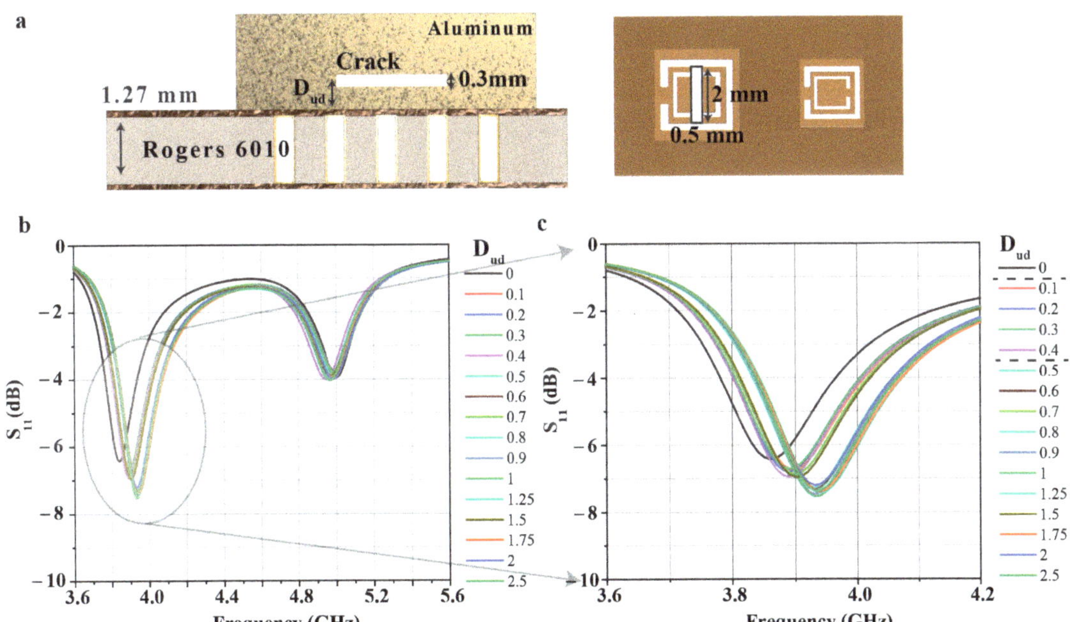

Figure 6. Simulation result of the proposed sensor for different depths from a metal surface: (**a**) side view and top view of structure and crack; (**b**,**c**) the simulated S_{11} for different crack positions: depth from a metal surface D_{ud} = 0–2.5 mm.

Figure 5 is the simulated result for checking the effect of the crack location deviation. In conclusion, since CSRR forms an electric field around the center of the ring, the frequency range varies depending on the location of the crack. First, Figure 5c,d show the simulated S11 of the proposed sensor for the *x*-direction position deviation of the crack. The original point (0, 0) is the intermediate position from the side of each CSRR. As shown in Figure 5b, when the crack moves from (−1, 0) to (8, 0) around CSRR2, the further away from (4, 0), the center of the CSRR, the smaller the range of frequency variation. In the same way, in Figure 5c, when the crack moves from (−9, 0) to (−2, 0) around CSRR1, the further away from (−5, 0), the center of the CSRR, the smaller the range of frequency variation. However, when the crack positions outside the CSRR where the electric field is weak, there is little change in frequency from the *x*-axis movement of the crack, as shown in Figure 5d,e. In addition, we checked the effect of the *y*-direction position deviation of the crack (Figure 5f). Similar to the change in the *x*-axis, it was shown that the closer the center of the CSRR, the greater the frequency change in the position deviation of the *y*-axis. For example, when the crack is in (4, 0), the frequency moves from 4.82 GHz to 5.18 GHz; when the crack is in (4, −1) or (4, 1), the frequency moves from 4.82 GHz to 5.1 GHz; and when the crack is in (4, −3) or (4, 3), the frequencies move from 4.82 GHz to 5 GHz. However, when the crack is outside the CSRR, there is little change in frequency from the *y*-axis movement of the crack.

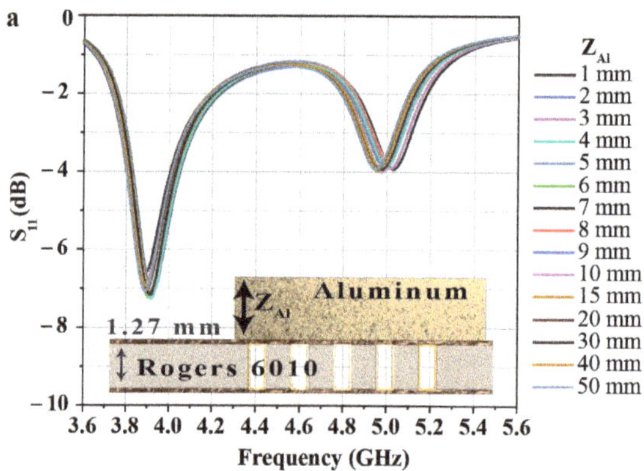

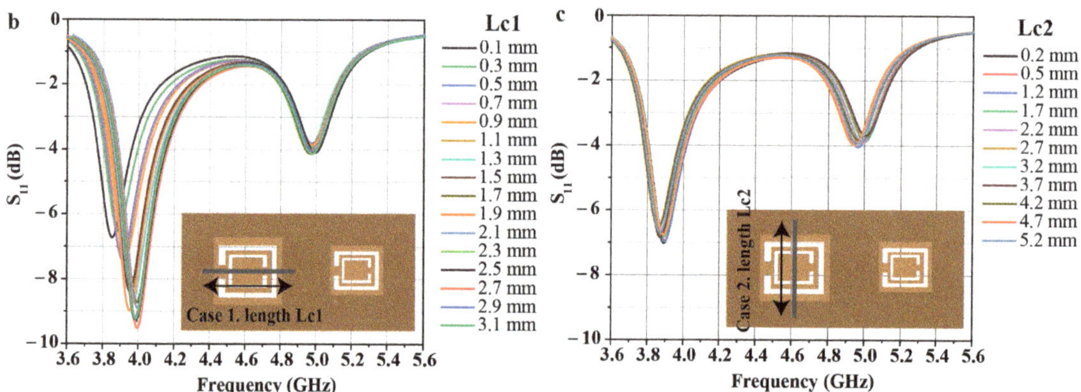

Figure 7. Simulation results of the proposed sensor for (**a**) different Al sheet thicknesses and (**b**,**c**) crack lengths.

Next, Figure 6 shows the detection capability of the crack under the metal surface. Generally, non-destructive testing can be used for detecting cracks under the surfaces. In the simulation, the crack of $0.5 \times 2 \times 0.3$ mm^3 is used, and the crack is positioned in the center of CSRR1, as shown in Figure 6a. Then, the crack is moved from the metal surface (D_{ud} = 0 mm) to the interior of the metal (D_{ud} = 2.5 mm). As a result, when D_{ud} changes from 0.1 mm to 0.4 mm, the frequency is changed from 3.82 GHz to 3.88 GHz, and when D_{ud} is bigger than 0.4 mm, the resonance frequency of the sensor generates at 3.94 GHz. In summary, the proposed sensor can detect whether a crack is on a metal surface or inside the metal, but it is difficult to distinguish how deep the crack penetrates accurately.

Finally, Figure 7 shows the effect of the thickness of the Al sheet for the proposed sensor and the detection capability for the crack length. First, the thickness of the Al sheet has little effect on the RF sensor because the metal material reflects almost all the waves on the surface, as shown in Figure 7a. Next, the length of the crack has a different effect on the RF sensor depending on the direction of the crack. Figure 7b shows that a change in frequency occurs when the crack is lengthened toward the gap of the ring since the electric field is strongly formed in the center of the CSRR and near the gap of the ring. Conversely, little frequency change appears when the crack is lengthened in a direction unrelated to the ring's gap, as shown in Figure 7c.

3. Fabrication of a Prototype Sensor and Measurement

3.1. Fabrication and Sensor Measurement Results

We fabricated a prototype SIW-based dual-crack sensor with various Al sheets in accordance with the proposed design. Figure 8a,b show top and bottom views of the fabricated sensor, and Figure 8c shows a side view of the fabricated sensor with the Al sheet. We employed four representative Al sheets with different straight crack widths and depths (Table 2); more complex shapes were not tested due to the difficulty of fabrication.

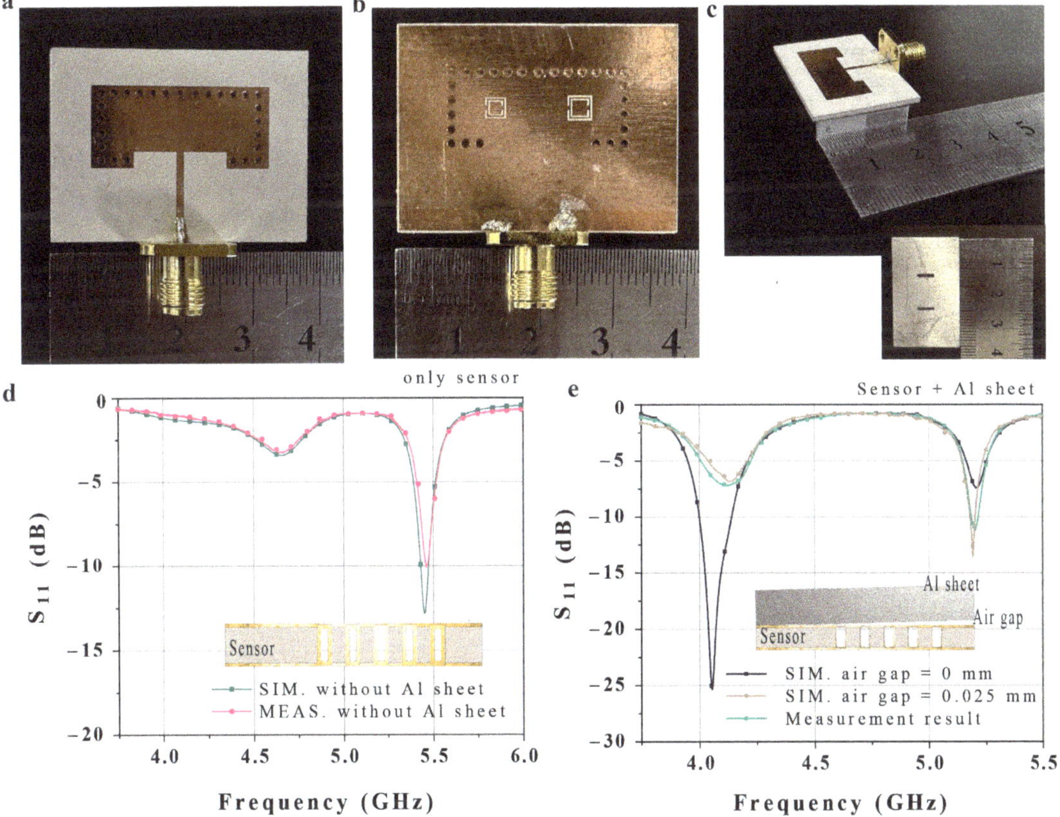

Figure 8. Fabricated prototype of the proposed dual-crack sensor: (**a**) top view, (**b**) bottom view, and (**c**) side view with the Al sheet. Simulation and measurement results for (**d**) the sensor with no Al sheet and (**e**) the sensor with an uncracked Al sheet.

Table 2. Crack sizes on the Al sheets.

Sheet	Crack Width (mm)	Crack Depth (mm)
1	0.85	0.3/0.5
2	0.95/1.05	0.1
3	0.95/1.05	0.3
4	0.95/1.05	0.5

The Al sheets were the same size as the detector (36 mm × 23 mm width and length) and were 5 mm thick. Because the cracks were positioned symmetrically, we measured eight situations using the four samples. The width and depth were changed, while the distance between the cracks was 10.8 mm, and the crack length was 6.1 mm. The fabricated cracks are located at the center of the CSRRs.

Figure 8c,d compares the simulation and measured sensor results for the sensor only and for the sensor with an uncracked aluminum sheet. The S-parameter of the fabricated sample is measured using a Keysight N5227B (Keysight, Santa Rosa, CA, USA) vector network analyzer. Figure 8c shows that the measurement results for the sensor without an Al sheet are consistent with the simulation results. However, although the sensor with an uncracked Al sheet exhibits a similar trend, there is some inconsistency with the simulation results, especially in terms of the magnitude of S11. This is ascribed to the presence of a small air gap between the prototype sensor and the Al sheet (Figure 8e). When we employed a 0.025 mm air gap between the sensor and Al sheet in the simulation, the results became consistent.

3.2. Crack Sensing Performance

Figure 9a presents the change in the resonant frequency due to the eight types of crack for each CSRR. In the absence of any cracks, the resonance frequency of the sensor is 5.20 GHz for CSRR1 and 4.12 GHz for CSRR2. In the presence of a crack with a width of 0.85 mm and a depth of 0.5 mm, the frequency increases to 4.32 GHz for CSRR2 and 5.34 GHz for CSRR1. In addition, as the width of the crack increases, the resonance frequency increases. On the other hand, when a crack has the same width, the resonance frequency increases as the depth of the crack reduces. Figure 9b,c present a more detailed plot of the measurement results for each CSRR, and the resonant frequency for each crack is summarized in Table 3.

Table 3. Proposed sensor performance for each CSRR.

	Crack Width × Depth (mm)							
	0.85 × 0.5	0.85 × 0.3	0.95 × 0.5	1.05 × 0.5	0.95 × 0.3	1.05 × 0.3	0.95 × 0.1	1.05 × 0.1
CSRR1 *	5.34	5.368	5.413	5.426	5.434	5.463	5.469	5.489
CSRR2 *	4.32	4.341	4.403	4.425	4.449	4.462	4.507	4.527

* Unit is GHz.

Figure 9b displays the measured S11 plotted against the CSRR2 resonant frequency for the different cracks. The resonant frequency ranges from 4.320 GHz for the 0.85 × 0.5 mm crack to 4.527 GHz for the 1.05 × 0.1 mm ^{2} crack. This demonstrates that the proposed sensor can detect cracks above CSRR2 at a resolution of 0.1 mm for the crack width and 0.2 mm for the crack depth. Figure 9c presents the measured S11 plotted against the CSRR1 resonant frequency for the different cracks. The resonant frequency ranges from 5.340 GHz for the 0.85 × 0.5 mm^2 crack to 5.489 GHz for the 1.05 × 0.1 mm^2 crack; thus, CSRR1 has the same sending resolution as CSRR2.

Figure 9d,e show the relationship between the resonant frequency and crack width and depth. Figure 9d presents the effect of the depth on the sensing of a crack with a fixed width of 0.95 and 1.05 mm at each CSRR. The average change in the resonant frequency is 20.7 and 18.3 MHz for CSRR1 and CSRR2, respectively, with a change of 0.1 mm in width. The fabricated sensor exhibits a linear reduction in the resonant frequency at both CSRRs with an increasing crack depth. Figure 6e presents the effect of the width on crack detection with a fixed depth of 0.3 and 0.5 mm at each CSRR. The average change in the resonant frequency is 28.0 and 33.7 MHz for CSRR1 and CSRR2, respectively, with a change of 0.2 mm in depth. The fabricated sensor thus exhibits a linear increase in the resonant frequency with an increase in the crack width.

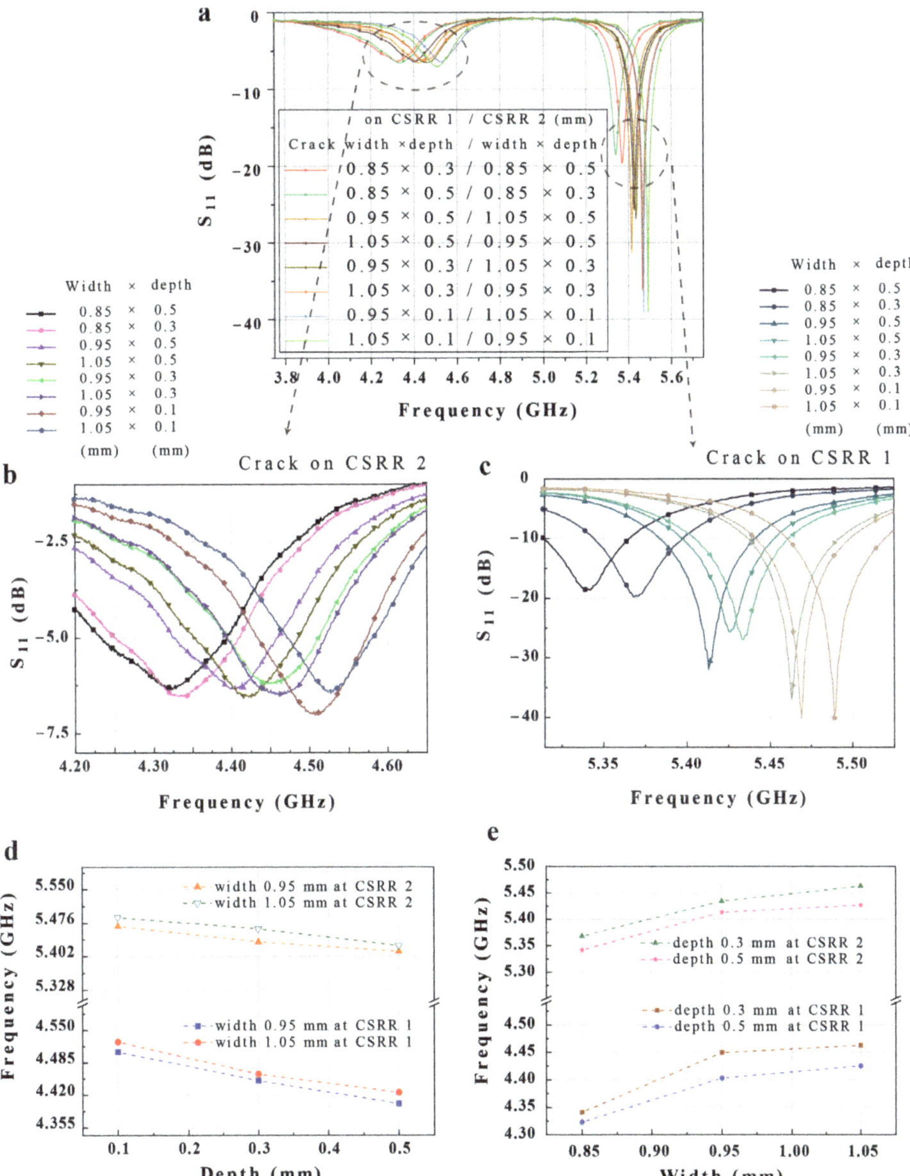

Figure 9. Measured S-parameters for the proposed dual-crack sensor (**a**) with various slot widths and depths at (**b**) CSRR 2 and (**c**) CSRR 1. Measured relationship between the frequency and various crack sizes: (**d**) depth and (**e**) width.

The proposed sensor experiences a decrease in the resonant frequency of 30.8 MHz when the crack width increases by 0.1 mm and an increase of 19.5 MHz when the crack depth increases by 0.2 mm. Thus, the proposed crack sensor can detect differences in crack width down to 0.1 mm and crack depth down to 0.2 mm.

The measured changes in the resonance due to changes in the crack dimension are generally linear, and the sensor can successfully simultaneously detect two different crack widths or depths, with each CSRR independent of the other. Cracks can thus be detected in situations where one crack is located near CSRR1, and another crack is located near CSRR2. In this respect, the measurement results exhibit the same trend as the simulation results and electromagnetic theory. The simulation results confirm that the effective capacitance falls with an increasing crack width; hence, the resonant frequency increases. In contrast, the effective capacitance increases with increasing crack depth, reducing the resonant frequency.

4. Conclusions

This study proposed a non-destructive multi-crack detection sensor consisting of differently sized CSRRs in a higher-mode SIW cavity to provide multi-resonant frequencies. Each CSRR on the SIW creates its own resonance, and the higher-mode SIW offers separated and concentrated E-fields for each CSRR. As a result, the proposed sensor has a high Q-factor and can simultaneously and independently detect multi-cracks. The reflective coefficient of the sensor is affected by the metal surface due to the strong external E-field from the CSRRs and SIW. When the metal has cracks, the efficient capacitance changes, which alters the resonant frequency of the nearby CSRR. Additionally, due to the high Q-factor, the proposed sensor can detect the metal crack at a high resolution. The effective capacitance of the sensor falls as the crack width increases, raising the resonant frequency, with the capacitance increasing as the crack depth becomes shallower, thus reducing the resonant frequency. The relationship between the resonance and the crack width or depth is almost linear. To demonstrate our proposed design, we designed and fabricated a dual-crack detection sensor using two independent CSRRs and a TE_{20}-mode SIW. The fabricated crack sensor distinguishes crack width differences down to 0.1 mm and crack depth differences to 0.2 mm and is able to two detect two differently sized cracks independently and simultaneously with significantly lower power consumption and scanning time. The proposed dual-crack detection sensor also exhibits a higher Q-factor than previously reported multi-detection sensors.

Author Contributions: Conceptualization, S.L.; methodology, Y.K., E.P., A.S. and S.L.; software, Y.K. and E.P.; validation, Y.K., E.P. and A.S.; formal analysis, Y.K.; investigation, Y.K., E.P. and S.L.; resources, Y.K. and E.P.; data curation, Y.K., E.P. and J.K.; writing—original draft preparation, Y.K. and E.P.; visualization, E.P. and J.K.; Supervision, S.L.; project administration, S.L.; funding acquisition, S.L. All authors have read and agreed to the published version of the manuscript.

Funding: This research was supported by a National Research Foundation (NRF) of Korea grant funded by the Korean government (MSIT) (2021R1A2C3005239 and 2021R1A4A2001316) and Chung-Ang University Research Scholarship Grants in 2022.

Data Availability Statement: Not applicable.

Conflicts of Interest: The authors declare no conflict of interest.

References

1. Kostromitin, K.I.; Khaliullin, R.S.; Skorobogatov, A.V. Using Destructive Testing Methods, X-ray Diffraction and Logic Analysis to Control Operation of Integral Circuits. In Proceedings of the 2020 International Conference on Industrial Engineering, Applications and Manufacturing (ICIEAM), Sochi, Russia, 18–22 May 2020; pp. 1–4.
2. Malek, J.; Kaouther, M. Destructive and non-destructive testing of concrete structures. *Jordan J. Civ. Eng.* **2014**, *8*, 432–441.
3. Wellander, N.; Elfsberg, M.; Sundberg, H.; Hurtig, T. Destructive testing of electronic components based on absorption cross section RC measurements. In Proceedings of the 2019 International Symposium on Electromagnetic Compatibility—EMC EUROPE, Barcelona, Spain, 2–6 September 2019; pp. 778–783.

4. Allwood, G.; Wild, G.; Hinckley, S. Optical Fiber Sensors in Physical Intrusion Detection Systems: A Review. *IEEE Sens. J.* **2016**, *16*, 5497–5509. [CrossRef]
5. Caizzone, S.; DiGiampaolo, E. Wireless Passive RFID Crack Width Sensor for Structural Health Monitoring. *IEEE Sens. J.* **2015**, *15*, 6767–6774. [CrossRef]
6. Chady, T.; Enokizono, M.; Sikora, R. Crack detection and recognition using an eddy current differential probe. *IEEE Trans. Magn.* **1999**, *35*, 1849–1852. [CrossRef]
7. Dutta, D.; Sohn, H.; Harries, K.A.; Rizzo, P. A Nonlinear Acoustic Technique for Crack Detection in Metallic Structures. *Struct. Health Monit.* **2009**, *8*, 251–262. [CrossRef]
8. Luo, D.; Yue, Y.; Li, P.; Ma, J.; Zhang, L.I.; Ibrahim, Z.; Ismail, Z. Concrete beam crack detection using tapered polymer optical fiber sensors. *Measurement* **2016**, *88*, 96–103. [CrossRef]
9. Tan, X.; Bao, Y. Measuring crack width using a distributed fiber optic sensor based on optical frequency domain reflectometry. *Measurement* **2021**, *172*, 108945. [CrossRef]
10. Wild, G.; Hinckley, S. Acousto-Ultrasonic Optical Fiber Sensors: Overview and State-of-the-Art. *IEEE Sens. J.* **2008**, *8*, 1184–1193. [CrossRef]
11. Zhang, Y. In Situ Fatigue Crack Detection using Piezoelectric Paint Sensor. *J. Intell. Mater. Syst. Struct.* **2006**, *17*, 843–852. [CrossRef]
12. Zhao, S.; Sun, L.; Gao, J.; Wang, J.; Shen, Y. Uniaxial ACFM detection system for metal crack size estimation using magnetic signature waveform analysis. *Measurement* **2020**, *164*, 108090. [CrossRef]
13. Ansari, M.A.H.; Jha, A.K.; Akhter, Z.; Akhtar, M.J. Multi-Band RF Planar Sensor Using Complementary Split Ring Resonator for Testing of Dielectric Materials. *IEEE Sens. J.* **2018**, *18*, 6596–6606. [CrossRef]
14. Dong, H.; Kang, W.; Liu, L.; Wei, K.; Xiong, J.; Tan, Q. Wireless passive sensor based on microstrip antenna for metal crack detection and characterization. *Meas. Sci. Technol.* **2019**, *30*, 045103. [CrossRef]
15. Huang, C.; Huang, B.; Zhang, B.; Li, Y.; Zhang, J.; Wang, K. An Electromagnetically Induced Transparency Inspired Antenna Sensor for Crack Monitoring. *IEEE Sens. J.* **2021**, *21*, 651–658. [CrossRef]
16. Javed, N.; Azam, M.A.; Amin, Y. Chipless RFID Multisensor for Temperature Sensing and Crack Monitoring in an IoT Environment. *IEEE Sens. Lett.* **2021**, *5*, 1–4. [CrossRef]
17. Khan, S.; Elbert, T.F. Complementary metaresonators based X-band hollow waveguide filter and crack detection sensor. In Proceedings of the 2017 11th European Conference on Antennas and Propagation (EUCAP), Paris, France, 19–24 March 2017; pp. 2346–2348.
18. Li, D.; Wang, Y. Thermally Stable Wireless Patch Antenna Sensor for Strain and Crack Sensing. *Sensors* **2020**, *20*, 3835. [CrossRef]
19. Martínez-Castro, R.E.; Jang, S.; Nicholas, J.; Bansal, R. Experimental assessment of an RFID-based crack sensor for steel structures. *Smart Mater. Struct.* **2017**, *26*, 085035. [CrossRef]
20. Zoughi, R.; Ganchev, S.I.; Huber, C. Measurement parameter optimization for surface crack detection in metals using an open-ended waveguide probe. In Proceedings of the Quality Measurement: The Indispensable Bridge between Theory and Reality (No Measurements? No Science! Joint Conference—1996: IEEE Instrumentation and Measurement Technology Conference and IMEKO Tec), Brussels, Belgium, 4–6 June 1996; pp. 1391–1394.
21. Bahrami, A.; Khouzani, M.K.; Mokhtari, S.A.; Zareh, S.; Mehr, M.Y. Root Cause Analysis of Surface Cracks in Heavy Steel Plates during the Hot Rolling Process. *Metals* **2019**, *9*, 801. [CrossRef]
22. Seo, D.C.; Lee, J.J.; Kwon, I.B. Monitoring of fatigue crack growth of cracked thick aluminum plate repaired with a bonded composite patch using transmission-type extrinsic Fabry–Perot interferometric optical fiber sensors. *Smart Mater. Struct.* **2002**, *11*, 917. [CrossRef]
23. Sharp, N.; Kuntz, A.; Brubaker, C.; Amos, S.; Gao, W.; Gupta, G. Crack detection sensor layout and bus configuration analysis. *Smart Mater. Struct.* **2014**, *23*, 055021. [CrossRef]
24. Abbasi, Z.; Baghelani, M.; Daneshmand, M. High-Resolution Chipless Tag RF Sensor. *IEEE Trans. Microw. Theory Tech.* **2020**, *68*, 4855–4864. [CrossRef]
25. Abdolrazzaghi, M.; Zarafi, M.H.; Floquet, C.F.A.; Daneshmand, M. Contactless Asphaltene Detection Using an Active Planar Microwave Resonator Sensor. *Energy Fuels* **2017**, *31*, 8784–8791. [CrossRef]
26. Cui, Y.; Ge, A. A Tunable High-Q Microwave Detector for On-Column Capillary Liquid Chromatography. *IEEE Trans. Instrum. Meas.* **2020**, *69*, 5978–5980. [CrossRef]
27. Zarifi, M.H.; Farsinezhad, S.; Shankar, K.; Daneshmand, M. Liquid Sensing Using Active Feedback Assisted Planar Microwave Resonator. *IEEE Microw. Wirel. Compon. Lett.* **2015**, *25*, 621–623. [CrossRef]
28. Mohammadi, S.; Adhikari, K.K.; Jain, M.C.; Zarifi, M.H. High-Resolution, Sensitivity-Enhanced Active Resonator Sensor Using Substrate-Embedded Channel for Characterizing Low-Concentration Liquid Mixtures. *IEEE Trans. Microw. Theory Tech.* **2022**, *70*, 576–586. [CrossRef]
29. Zarifi, M.H.; Thundat, T.; Daneshmand, M. High resolution microwave microstrip resonator for sensing applications. *Sens. Actuators A Phys.* **2015**, *233*, 224–230. [CrossRef]
30. Ahmad, S.; Ghaffar, A.; Hussain, N.; Kim, N. Compact Dual-Band Antenna with Paired L-Shape Slots for On- and Off-Body Wireless Communication. *Sensors* **2021**, *21*, 7953. [CrossRef] [PubMed]

31. Bakkali, A.; Pelegri-Sebastia, J.; Sogorb, T.; Llario, V.; Bou-Escriva, A. A Dual-Band Antenna for RF Energy Harvesting Systems in Wireless Sensor Networks. *J. Sens.* **2016**, *2016*, 5725836. [CrossRef]
32. Park, E.; Lim, D.; Lim, S. Dual-Band Band-Pass Filter with Fixed Low Band and Fluidically-Tunable High Band. *Sensors* **2017**, *17*, 1884. [CrossRef]
33. Zhang, J.; Meng, J.; Li, W.; Yan, S.; Vandenbosch, G.A.E. A Wearable Button Antenna Sensor for Dual-Mode Wireless Information and Power Transfer. *Sensors* **2021**, *21*, 5678. [CrossRef]
34. Zhu, L.; Farhat, M.; Chen, Y.C.; Chen, P.Y. A Compact, Passive Frequency-Hopping Harmonic Sensor Based on a Microfluidic Reconfigurable Dual-Band Antenna. *IEEE Sens. J.* **2020**, *20*, 12495–12503. [CrossRef]
35. Albishi, A.M.; Boybay, M.S.; Ramahi, O.M. Complementary Split-Ring Resonator for Crack Detection in Metallic Surfaces. *IEEE Microw. Wirel. Compon. Lett.* **2012**, *22*, 330–332. [CrossRef]
36. Chen, Z.; Lin, X.Q.; Yan, Y.H.; Xiao, F.; Khan, M.T.; Zhang, S. Noncontact Group-Delay-Based Sensor for Metal Deformation and Crack Detection. *IEEE Trans. Ind. Electron.* **2021**, *68*, 7613–7619. [CrossRef]
37. Ebrahimi, A.; Scott, J.; Ghorbani, K. Compact Dual Bandpass Filter Using Dual-Split Ring Resonator for 5G Upper Microwave Flexible Use Services. *IEEE Sens. J.* **2020**, *20*, 185–192. [CrossRef]
38. Sowjanya, A.; Vakula, D. Compact Dual-Mode Wideband Filter Based on Complementary Split-Ring Resonator. *J. Electromagn. Eng. Sci.* **2022**, *22*, 434–439. [CrossRef]
39. Kiani, S.; Rezaei, P.; Navaei, M. Dual-sensing and dual-frequency microwave SRR sensor for liquid samples permittivity detection. *Measurement* **2020**, *160*, 107805. [CrossRef]
40. Marindra, A.M.J.; Sutthaweekul, R.; Tian, G.Y. Depolarizing Chipless RFID Sensor Tag for Characterization of Metal Cracks Based on Dual Resonance Features. In Proceedings of the 2018 10th International Conference on Information Technology and Electrical Engineering (ICITEE), Bali, Indonesia, 24–26 July 2018; pp. 73–78.
41. Jang, S.-Y.; Yang, J.-R. Double Split-Ring Resonator for Dielectric Constant Measurement of Solids and Liquids. *J. Electromagn. Eng. Sci.* **2022**, *22*, 122–128. [CrossRef]
42. Salim, A.; Naqvi, A.H.; Pham, A.D.; Lim, S. Complementary Split-Ring Resonator (CSRR)-Loaded Sensor Array to Detect Multiple Cracks: Shapes, Sizes, and Positions on Metallic Surface. *IEEE Access* **2020**, *8*, 151804–151816. [CrossRef]
43. Samad, A.; Hu, W.D.; Shahzad, W.; Lightart, L.P.; Raza, H. High-Precision Complementary Metamaterial Sensor Using Slotted Microstrip Transmission Line. *J. Sens.* **2021**, *2021*, 8888187. [CrossRef]
44. Su, P.; Yang, X.; Wang, J.; Wang, Z.; Peng, H. Detection of Metal Surface Cracks Based on Liquid Switch Controlled Spoof Surface Plasmon Polaritons. *IEEE Sens. J.* **2022**, *22*, 1287–1294. [CrossRef]
45. Wei, F.; Qin, P.Y.; Guo, Y.J.; Ding, C.; Shi, X.W. Compact Balanced Dual- and Tri-Band BPFs Based on Coupled Complementary Split-Ring Resonators (C-CSRR). *IEEE Microw. Wirel. Compon. Lett.* **2016**, *26*, 107–109. [CrossRef]
46. Mohammadi Shirkolaei, M.; Ghalibafan, J. Unbalanced CRLH Behavior of Ferrite-Loaded Waveguide Operated below Cutoff Frequency. *Waves Random Complex Media* **2022**, *32*, 755–770. [CrossRef]
47. Mohammadi Shirkolaei, M.; Ghalibafan, J. Magnetically Scannable Slotted Waveguide Antenna Based on the Ferrite with Gain Enhancement. *Waves Random Complex Media* **2021**, 1–11. [CrossRef]
48. Shahzad, W.; Hu, W.; Ali, Q.; Raza, H.; Abbas, S.M.; Ligthart, L.P. A Low-Cost Metamaterial Sensor Based on DS-CSRR for Material Characterization Applications. *Sensors* **2022**, *22*, 2000. [CrossRef] [PubMed]
49. Yi, X.; Cho, C.; Cooper, J.; Wang, Y.; Tentzeris, M.M.; Leon, R.T. Passive wireless antenna sensor for strain and crack sensing—Electromagnetic modeling, simulation, and testing. *Smart Mater. Struct.* **2013**, *22*, 085009. [CrossRef]
50. Yun, T.; Lim, S. High-Q and miniaturized complementary split ring resonator-loaded substrate integrated waveguide microwave sensor for crack detection in metallic materials. *Sens. Actuators A Phys.* **2014**, *214*, 25–30. [CrossRef]
51. Armghan, A.; Alanazi, T.M.; Altaf, A.; Haq, T. Characterization of Dielectric Substrates Using Dual Band Microwave Sensor. *IEEE Access* **2021**, *9*, 62779–62787. [CrossRef]
52. Falcone, F.; Lopetegi, T.; Laso, M.A.G.; Baena, J.D.; Bonache, J.; Beruete, M.; Marques, R.; Martin, F.; Sorolla, M. Babinet principle applied to the design of metasurfaces and metamaterials. *Phys. Rev. Lett.* **2004**, *93*, 197401. [CrossRef]
53. Pendry, J.B.; Holden, A.J.; Robbins, D.J.; Stewart, W.J. Magnetism from conductors and enhanced nonlinear phenomena. *IEEE Trans. Microw. Theory Tech.* **1999**, *47*, 2075–2084. [CrossRef]
54. Albishi, A.M.; Mirjahanmardi, S.H.; Ali, A.M.; Nayyeri, V.; Wasly, S.M.; Ramahi, O.M. Intelligent Sensing Using Multiple Sensors for Material Characterization. *Sensors* **2019**, *19*, 4766. [CrossRef]
55. Albishi, A.M. A Novel Coupling Mechanism for CSRRs as Near-Field Dielectric Sensors. *Sensors* **2022**, *22*, 3313. [CrossRef]
56. Albishi, A.M.; Badawe, M.K.E.; Nayyeri, V.; Ramahi, O.M. Enhancing the Sensitivity of Dielectric Sensors With Multiple Coupled Complementary Split-Ring Resonators. *IEEE Trans. Microw. Theory Tech.* **2020**, *68*, 4340–4347. [CrossRef]
57. Albishi, A.M.; Ramahi, O.M. Surface crack detection in metallic materials using sensitive microwave-based sensors. In Proceedings of the 2016 IEEE 17th Annual Wireless and Microwave Technology Conference (WAMICON), Clearwater, FL, USA, 11–13 April 2016; pp. 1–3.
58. Haq, T.; Ruan, C.; Ullah, S.; Fahad, A.K. Dual Notch Microwave Sensors Based on Complementary Metamaterial Resonators. *IEEE Access* **2019**, *7*, 153489–153498. [CrossRef]
59. Liu, W.; Zhang, J.; Huang, K. Dual-Band Microwave Sensor Based on Planar Rectangular Cavity Loaded With Pairs of Improved Resonator for Differential Sensing Applications. *IEEE Trans. Instrum. Meas.* **2021**, *70*, 1–8. [CrossRef]

60. Nogueira, J.K.A.; Oliveira, J.G.D.; Paiva, S.B.; Neto, V.P.S.; D'Assunção, A.G. A Compact CSRR-Based Sensor for Characterization of the Complex Permittivity of Dielectric Materials. *Electronics* **2022**, *11*, 1787. [CrossRef]
61. Omer, A.E.; Shaker, G.; Safavi-Naeini, S.; Kokabi, H.; Alquié, G.; Deshours, F.; Shubair, R.M. Low-cost portable microwave sensor for non-invasive monitoring of blood glucose level: Novel design utilizing a four-cell CSRR hexagonal configuration. *Sci. Rep.* **2020**, *10*, 15200. [CrossRef]
62. Omer, A.F.; Shaker, G.; Safavi-Naeini, S.; Ngo, K.; Shubair, R.M.; Alquié, G.; Deshours, F.; Kokabi, H. Multiple-Cell Microfluidic Dielectric Resonator for Liquid Sensing Applications. *IEEE Sens. J.* **2021**, *21*, 6094–6104. [CrossRef]
63. Wu, W.J.; Zhao, W.S.; Wang, D.W.; Yuan, B.; Wang, G. Ultrahigh-Sensitivity Microwave Microfluidic Sensors Based on Modified Complementary Electric-LC and Split-Ring Resonator Structures. *IEEE Sens. J.* **2021**, *21*, 18756–18763. [CrossRef]
64. Chen, C.M.; Xu, J.; Yao, Y. Fabrication of miniaturized CSRR-loaded HMSIW humidity sensors with high sensitivity and ultra-low humidity hysteresis. *Sens. Actuators B Chem.* **2018**, *256*, 1100–1106. [CrossRef]
65. Zeng, X.; Bi, X.; Cao, Z.; Wan, R.; Xu, Q. High Selectivity Dual-Wideband Balun Filter Utilizing a Multimode T-line Loaded Middle-Shorted CSRR. *IEEE Trans. Circuits Syst. II Express Briefs* **2020**, *67*, 2447–2451. [CrossRef]
66. Albishi, A.; Ramahi, O. Ultrasensitive microwave near-field based sensors for crack detection in metallic materials. In Proceedings of the 2016 IEEE International Symposium on Antennas and Propagation (APSURSI), Fajardo, PR, USA, 26 June–1 July 2016; pp. 1955–1956.
67. Kiani, S.; Rezaei, P.; Navaei, M.; Abrishamian, M.S. Microwave Sensor for Detection of Solid Material Permittivity in Single/Multilayer Samples With High Quality Factor. *IEEE Sens. J.* **2018**, *18*, 9971–9977. [CrossRef]
68. Hao, H.; Wang, D.; Wang, Z. Design of Substrate-Integrated Waveguide Loading Multiple Complementary Open Resonant Rings (CSRRs) for Dielectric Constant Measurement. *Sensors* **2020**, *20*, 857. [CrossRef] [PubMed]
69. Eom, S.; Memon, M.U.; Lim, S. Frequency-Switchable Microfluidic CSRR-Loaded QMSIW Band-Pass Filter Using a Liquid Metal Alloy. *Sensors* **2017**, *17*, 699. [CrossRef] [PubMed]
70. Kou, H.; Tan, Q.; Wang, Y.; Zhang, G.; Shujing, S.; Xiong, J. A microwave SIW sensor loaded with CSRR for wireless pressure detection in high-temperature environments. *J. Phys. D Appl. Phys.* **2019**, *53*, 085101. [CrossRef]
71. Liu, Z.; Xiao, G.; Zhu, L. Triple-Mode Bandpass Filters on CSRR-Loaded Substrate Integrated Waveguide Cavities. *IEEE Trans. Compon. Packag. Manuf. Technol.* **2016**, *6*, 1099–1105. [CrossRef]
72. Zhang, H.; Kang, W.; Wu, W. Dual-band substrate integrated waveguide bandpass filter utilising complementary split-ring resonators. *Electron. Lett.* **2018**, *54*, 85–87. [CrossRef]
73. Zhang, Q.L.; Yin, W.Y.; He, S.; Wu, L.S. Compact Substrate Integrated Waveguide (SIW) Bandpass Filter with Complementary Split-Ring Resonators (CSRRs). *IEEE Microw. Wirel. Compon. Lett.* **2010**, *20*, 426–428. [CrossRef]
74. Mohammadi, S.M.; Aslinezhad, M. Substrate integrated waveguide filter based on the magnetized ferrite with tunable capability. *Microw. Opt. Technol. Lett.* **2021**, *63*, 1120–1125.
75. Dong, Y.D.; Tao, Y.; Tatsuo, I. Substrate integrated waveguide loaded by complementary split-ring resonators and its applications to miniaturized waveguide filters. *IEEE Trans. Microw. Theory Tech.* **2009**, *57*, 2211–2223. [CrossRef]
76. Wang, Q.; Bi, K.; Hao, Y.; Guo, L.; Dong, G.; Wu, H.; Lei, M. High-Sensitivity Dielectric Resonator-Based Waveguide Sensor for Crack Detection on Metallic Surfaces. *IEEE Sens. J.* **2019**, *19*, 5470–5474. [CrossRef]
77. Rajni; Kaur, A.; Marwaha, A. Crack Detection on Metal Surfaces with an Array of Complementary Split Ring Resonators. *Int. J. Comput. Appl.* **2015**, *119*, 16–19.
78. Baena, J.D.; Bonache, J.; Martin, F.; Sillero, R.M.; Falcone, F.; Lopetegi, T.; Laso, M.A.G.; Garcia-Garcia, J.; Gil, I.; Portillo, M.F.; et al. Equivalent-circuit models for split-ring resonators and complementary split-ring resonators coupled to planar transmission lines. *IEEE Trans. Microw. Theory Tech.* **2005**, *53*, 1451–1461. [CrossRef]
79. Memon, M.U.; Lim, S. Review of Electromagnetic-Based Crack Sensors for Metallic Materials (Recent Research and Future Perspectives). *Metals* **2016**, *6*, 172. [CrossRef]
80. Albishi, A.M.; Ramahi, O.M. Microwaves-Based High Sensitivity Sensors for Crack Detection in Metallic Materials. *IEEE Trans. Microw. Theory Tech.* **2017**, *65*, 1864–1872. [CrossRef]
81. Chuma, E.L.; Iano, Y.; Fontgalland, G.; Roger, L.L.B. Microwave Sensor for Liquid Dielectric Characterization Based on Metamaterial Complementary Split Ring Resonator. *IEEE Sens. J.* **2018**, *18*, 9978–9983. [CrossRef]
82. Cassivi, Y.; Wu, K. Low cost microwave oscillator using substrate integrated waveguide cavity. *IEEE Microw. Wirel. Compon. Lett.* **2003**, *13*, 48–50. [CrossRef]
83. Cui, T.J.; Lin, X.Q.; Cheng, Q.; Ma, H.F.; Yang, X.M. Experiments on evanescent-wave amplification and transmission using metamaterial structures. *Phys. Rev. B* **2006**, *73*, 245119. [CrossRef]

 *micromachines*

Article

3D Numerical Simulation and Structural Optimization for a MEMS Skin Friction Sensor in Hypersonic Flow

Huihui Guo [1,2], **Xiong Wang** [3,*], **Tingting Liu** [1,*], **Zhijiang Guo** [1] **and Yang Gao** [1]

1 School of Information Engineering, Southwest University of Science and Technology, Mianyang 621010, China
2 Robot Technology Used for Special Environment Key Laboratory of Sichuan Province, Mianyang 621010, China
3 Hypervelocity Aerodynamics Institute, China Aerodynamics Research and Development Center, Mianyang 621010, China
* Correspondence: wangx81@cardc.cn (X.W.); liutingting@swust.edu.cn (T.L.); Tel.: +86-816-2465206 (X.W.); +86-816-6089322 (T.L.)

Abstract: The skin friction of a hypersonic vehicle surface can account for up to 50% of the total resistance, directly affecting the vehicle's effective range and load. A wind tunnel experiment is an important and effective method to optimize the aerodynamic shape of aircraft, and Micro-Electromechanical System (MEMS) skin friction sensors are considered the promising sensors in hypersonic wind tunnel experiments, owing to their miniature size, high sensitivity, and stability. However, the sensitive structure including structural appearance, a gap with the package shell, and flatness of the sensor will change the measured flow field and cause the accurate measurement of friction resistance. Aiming at the influence of sensor-sensitive structure on wall-flow characteristics and friction measurement accuracy, the two-dimensional and three-dimensional numerical models of the sensor in the hypersonic flow field based on Computational Fluid Dynamics (CFD) are presented respectively in this work. The model of the sensor is verified by using the Blathius solution of two-dimensional laminar flow on a flat plate. The results show that the sensor model is in good agreement with the Blathius solution, and the error is less than 0.4%. Then, the influence rules of the sensitive structure of the sensor on friction measurement accuracy under turbulent flow and laminar flow conditions are systematically analyzed using 3D numerical models of the sensor, respectively. Finally, the sensor-sensitive unit structure's design criterion is obtained to improve skin friction's measurement accuracy.

Keywords: skin friction sensor; 3D models; CFD; turbulent flow; laminar flow; friction measurement accuracy

Citation: Guo, H.; Wang, X.; Liu, T.; Guo, Z.; Gao, Y. 3D Numerical Simulation and Structural Optimization for a MEMS Skin Friction Sensor in Hypersonic Flow. *Micromachines* **2022**, *13*, 1487. https://doi.org/10.3390/mi13091487

Academic Editors: Yong Ruan, Weidong Wang and Zai-Fa Zhou

Received: 18 July 2022
Accepted: 5 September 2022
Published: 7 September 2022

Publisher's Note: MDPI stays neutral with regard to jurisdictional claims in published maps and institutional affiliations.

1. Introduction

The skin friction of an aircraft surface is the main part of its total resistance, which can account for up to 50% and greatly limits the effective range and load of hypersonic vehicles. A wind tunnel experiment is an important way and effective method to optimize the aerodynamic shape of aircraft. Therefore, the skin friction measurement of the aircraft model is an important basic item in wind tunnel experiment research. In the hypersonic wind tunnel experiment, obtaining the accurate friction distribution of the model points or surfaces puts forward higher requirements for the size of the sensor limited by the volume of the aircraft model. In recent years, several types of MEMS sensors have been reported to measure skin friction including capacitance type and comb differential capacitance type [1–3], piezoresistive type [4–6], and piezoelectric type [7,8]. Those MEMS sensors have similar characteristics such as small size, high sensitivity, and high resolution in a small measurement range (several Pa).

Owing to their miniature size, high sensitivity, and stability, the MEMS skin friction sensors are considered the promising sensors in wind tunnel experiments. For example,

Jiang et al. [1] have reported a supporting cantilever and plate differential capacitive skin friction sensor for low-speed wind tunnels with high sensitivity and a minimum detection limit of 0.05 Pa in a range from 0.1–2 Pa. Von P et al. [7] have reported a wall shear stress sensor using four piezoresistors as a transducer in the cantilever, with a resolution of 0.01 Pa in the range 0–2 Pa. T Kim T et al. [9] have reported a piezoelectric floating element shear stress sensor for the wind tunnel flow measurement with the high sensitivity 56.5 pc/Pa. However, these sensors are mainly used for skin friction detection in low wind tunnels because their sensing element such as comb capacitance was exposed in the flow field.

With the development of supersonic aircraft, direct skin friction measurements under hypersonic conditions are strongly required in wind tunnel experiments. At present, the environment of the hypersonic wind tunnel is relatively harsh, such as heavy load, particle inclusion in gas, etc. It is difficult for various comb capacitive sensors to survive in the hypersonic wind tunnel. To adapt to the harsh measurement environment and large normal loads in hypersonic fields, a MEMS skin friction sensor has been developed in our previous work [3,10], which adopts the floating element even with the measured wall and the signal output micro-structure being isolated from the hypersonic field. The sensor is composed of a silicon differential-capacitor embedded with a floating element, signal readout circuit, and a package metal shell, which was fabricated by using various micro-mechanical processes and micro-assembly technology. Experiments result show that the MEMS has good linearity, small size, high sensitivity, and repeatability of static calibration, the resolution is 0.1 Pa in ranges from 0 to 100 Pa, and the repeatability accuracy and linearity are better than 1%. However, the deviations between the measured skin friction coefficients and the analytical values are close to 100% under laminar flow conditions [10]. The reason for the large measurement error under the condition of laminar flow is that the friction coefficient itself is very small, and the small flow state interference from the sensor structure will cause a large deviation.

Generally, the height of the vicious bottom layer of low-speed and low shear flow is usually greater than 100 microns, and the influence of sensor structure error (tens of microns) on the fluid-structure can be ignored; the fluid-structure is considered hydraulic smoothing [10,11]. However, in high-speed and high shear flow, especially laminar flow, the height of the vicious bottom layer may be in the order of microns. At this time, the fluid structure is not considered hydraulic smoothing, the structure error of the sensor including the protrusion of the sensing element and the measurement gap will change the characteristics of the fluid, which will seriously affect the accuracy of the sensor in measuring the skin friction resistance. Therefore, it is of great significance to study the influence of sensor structure on the high-speed flow field. In this work, the two-dimensional and three-dimensional numerical models of the sensor in plate model under the hypersonic flow field based on CFD are presented respectively to research the influence of sensor-sensitive structure on wall-flow characteristics and friction perceptual characteristics. In addition, to improve the measurement accuracy of skin friction, the design criterion of the sensor-sensitive unit structure should be obtained.

2. A 3D Flat Plate Model Embedded with a MEMS Sensor

In the wind tunnel experiment, the measurement method of wall friction resistance is mainly based on the conventional friction balance and the friction measurement technology of the MEMS friction sensor, which is based on the plate model. The flat plate model is a classical and simple model, which is conducive to the installation of the sensor and the observation of the flow characteristics of the surface, and is also convenient for comparison with the Bratisius solution. The model of wind tunnel validation test is a flat plate model with tail support, and the MEMS sensors are installed in the inner cavity of the flat plate model, and the sensing surface of the measuring head is flush with the surface to be measured of the test model to directly feel the skin friction as shown in Figure 1a. The total length of the plate model is 400 mm and the width is 120 mm. The front sensor is 147 mm from the front edge of the plate model, and the rear sensor is 130 mm from the rear end face of the plate model. The two sides are symmetrically distributed along the neutral plane of the plate model, with a distance of 80 mm. The structure of MEMS skin friction is composed of a silicon differential-capacitor embedded with a floating element, signal readout circuit, and package metal shell as shown in Figure 1b. This paper focuses on the influence of the structure of the MEMS sensor on the measured flow field in hypersonic; therefore, except for the structure of the floating element, other structural parameters of the sensor remain unchanged. The structure parameters of the sensor are described in detail in our previous work [9]. The working principle of this sensor is explained as the skin friction causes the torsional deflection of clamped–clamped elastic beams by the floating element, and the torsional deflection is transferred into a capacitance variation by one pair of differential capacitors (as shown in Figure 1b). Finally, the friction coefficient is measured using the signal output circuit by measuring the change of the capacitors.

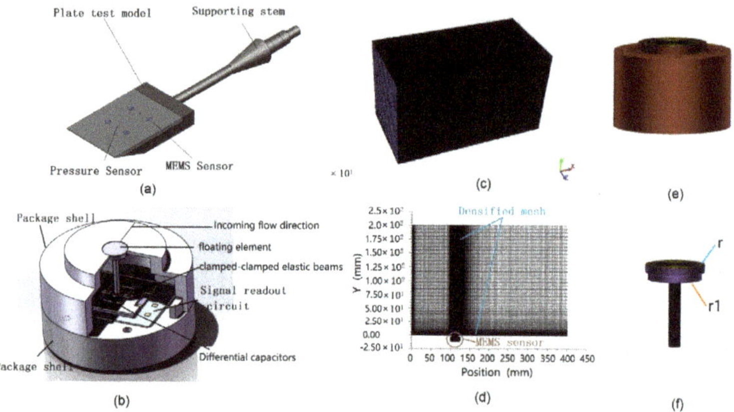

Figure 1. (**a**) Schematic of a flat plate model embedded with MEMS sensors: (**b**) schematic of the MEMS skin friction sensor; (**c**) the 3D mesh model of the flat plate model; (**d**) the lateral section of the 3D model as 2D plate model of the flat plate; (**e**) the 3D mesh model of the sensor; (**f**) the 3D mesh model of the floating element.

To research the influence of sensor-sensitive structure on wall-flow characteristics and friction perceptual characteristics, a 3D flat plate test model embedded with MEMS sensors is built by fluent which is a computer program used to simulate fluid flow and heat conduction with complex shapes, and the feature dimensions of the 3D model are 400 mm ×120 mm × 200 mm. The 2D plate model is the lateral section of the three-dimensional model, which is mainly used to compare with the Blasius solution to verify the effectiveness of the calculation model. In addition, the near-wall surface mesh of the numerical model needs to be densified to obtain more accurate numerical results, and the grid spacing of the first layer of the wall is 0.001 mm as shown in Figure 1d. The 3D model of the MEMS sensor is shown in Figure 1e, in which the diameters of the floating element with a two-layer structure are 5 mm (r) and 0.45 mm (r_1) respectively, and the thickness is all 0.9 mm as shown in Figure 1f. Different from the previous structure [9], the floating element structure in this paper is large in the upper layer and small in the lower layer.

Combined with the flow field parameters of the wind tunnel test, the initial condition of the inner cavity of the friction sensor is set to the vacuum state, and the boundary conditions include wall boundary conditions, inflow boundary conditions and outflow boundary conditions are determined as follows: (1) the measured wall boundary meets the requirements of a no-slip wall ($V = 0$) for hypersonic viscous flow, and the wall temperature is constant temperature wall ($T_w = 300$ K); (2) inlet boundary of incoming flow is set as far-field input of pressure, the simulated incoming flow conditions are shown in Table 1; (3) outlet boundary selection pressure outlet boundary condition. Then, the numerical simulation of the $0°$ angle of attack attitude of the 2D plate model is calculated. The wind tunnel flow field parameters (such as dynamic pressure q_∞, static pressure P_∞, and unit Reynolds number R_e/L) can be calculated according to the incoming total pressure P_0, total temperature T_0, and incoming Mach number M_∞ as follows:

$$P_\infty = P_0(1 + 0.2M_\infty^2)^{-3.5} \tag{1}$$

$$T_\infty = T_0(1 + 0.2M_\infty^2)^{-1} \tag{2}$$

$$q_\infty = 0.7P_\infty M_\infty^2 \tag{3}$$

$$\mu_\infty = 1.458 \times 10^{-6} \frac{T_\infty^{1.5}}{T_\infty + 110.4} \tag{4}$$

$$R_e/L = \frac{0.05902M_\infty}{(1 + 0.2M_\infty^2)^3} \cdot \frac{P_0}{\mu_\infty} \cdot \left(\frac{1.4}{T_0}\right)^{0.5} \tag{5}$$

Table 1. The main parameters of the wind tunnel.

Nominal Mach Number (M_∞)	Total Pressure P_0 (M Pa)	Total Temperature T_0 (K)	Static Pressure P_∞ (Pa)	Dynamic Pressure q_∞ (Pa)	Unit Reynolds Number Re/L (L/m)
	0.5	417	316.68	7980.4	5.679×10^6
	1.0	437	633.33	15961	1.053×10^7
6	1.5	449	950.04	23941	1.506×10^7
	2.0	458	1266.7	31924	1.941×10^7
	2.5	466	1583.4	39902	2.361×10^7

The calculation of friction coefficient (C_f) is as follows:

$$C_f = \frac{\tau_w}{q_\infty} \tag{6}$$

where τ_w is the measured skin friction, and q_∞ is the dynamic pressure.

The 3D calculation method is verified by using the Blasius solution of two-dimensional plate laminar flow under the condition of total incoming flow pressure of 1.5 MPa. In this work, the 2D plate model is the lateral section of the 3D model, which is mainly used to compare with the Blasius solution using the laminar flow model and turbulent model (k-e model). The 2D streamline diagram in the friction sensor is shown in Figure 2a, the X direction is the incoming flow direction, and the surface streamline shows that there is suction in the moving gap. It also can be seen that the airflow enters the cavity of the friction sensor along the wall from the gap inlet causing pressure resistance to the leading edge of the floating element head and affecting the friction coefficient. Meanwhile, the numerical simulation value of the friction coefficient at the installation position of the friction sensor (x = 112.6 mm) is 5.09×10^{-4} (directly read out in the model). Compared with Blasius solution 5.11×10^{-4} ($C_f = \frac{0.667}{\sqrt{Rex}}$), the difference is about 0.39%. Hence, the calculated results of plate surface friction coefficient are in good agreement with the Blasius solution using the laminar flow model as shown in Figure 2b, which also shows that the numerical simulation method and calculation model constructed in this paper is effective.

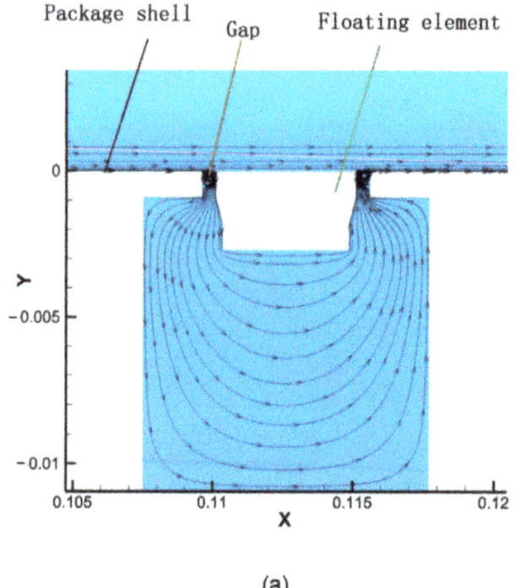

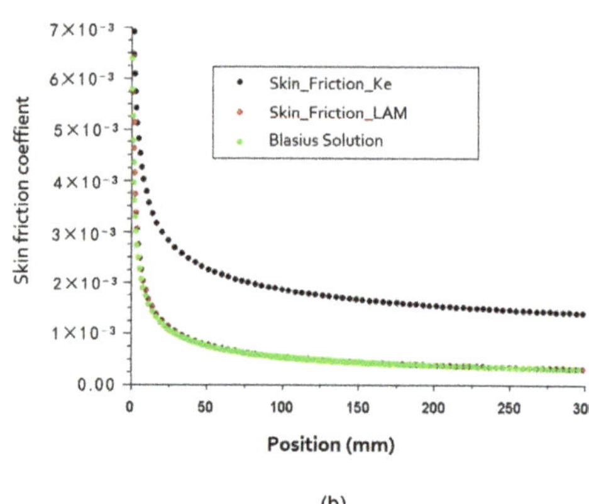

Figure 2. (**a**) The 2D streamline diagram in the friction sensor; (**b**) the skin friction coefficient of the flat plate in the laminar flow model and turbulent model under a total pressure of 1.5 MPa incoming flows.

In addition, as shown in Figure 2b, it can also be seen that the friction coefficient in the laminar flow model state is smaller than that in the turbulent state, and the variation trend of the friction coefficient under the turbulent state is similar to that under laminar state, but the friction coefficient is larger than that under laminar state, and its value is around 0.0017.

3. Structural Optimization of the Floating Element

In our previous wind tunnel experiments, the deviations between the measured skin friction coefficients and the analytical values are closed to 100% under laminar flow conditions [10]. The main reason for the error is that the shape of the floating element has an obvious influence on the flow disturbance of the wall fluid, which has an important impact on the friction measurement error. The velocity cloud diagram in the front direction of the floating element of the central section of the friction sensor in the direction of the incoming flow is shown in Figure 3b. It can be seen that the flow field on the surface of the floating element changes due to the suction effect of the moving gap, and the blocking effect of the cylindrical floating element and the diversion effect of the annular gap cause the lateral flow around the cylindrical floating element, and the friction coefficient at the edge of the floating element increases significantly as shown in Figure 3c.

In order to reduce the influence of the sensor structure on the flow disturbance of the measured wall fluid, the improved floating element is shown in Figure 3d. In addition, the 3D sensor model is also built using CFD, and the velocity distributions in the X direction and Y direction are obtained under a total pressure of 1.5 MPa as shown in Figure 3e,f. The results show that the improved floating element structure can reduce the lateral flow around the cylindrical floating element. A three-layer floating element structure (as shown in Figure 4b) model was also established to further study the influence of floating element structure on the wall flow field and friction measurement error.

The structural error of the prepared sensor mainly includes the gap, misalignment, and non-coaxial between the floating element and the package as shown in Figure 4a. The coaxial assembly problem of the sensor has been solved based on visual alignment technology. The coaxiality measurement of the MEMS sensor is shown in Figure 4c. The position error is only 2 μm, which is about 0.4% of the moving gap with 500 μm. However, the misalignment is difficult to control because the machining accuracy of parts is less than 10 microns. Combined with the wind tunnel experiments results [10] and error analysis results [6,12], the structure appearance of the floating element, the misalignment between the floating element and the package shell, and the gap of the sensor is the main source of measurement error.

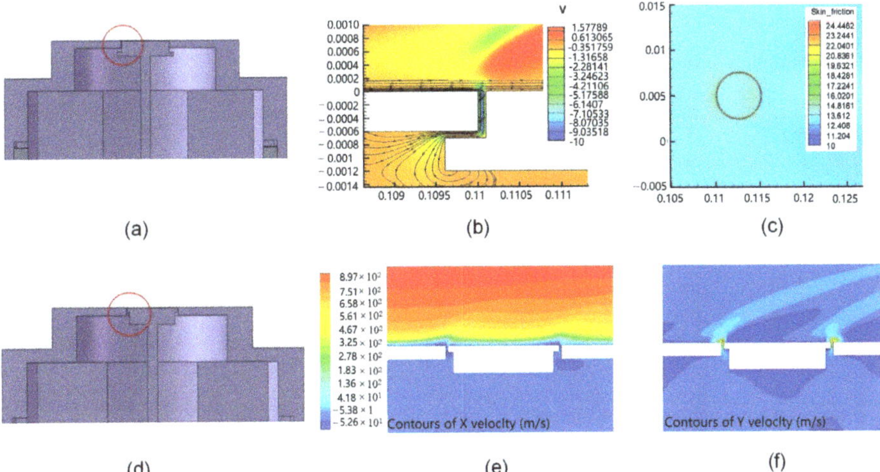

Figure 3. (**a**) The structure of the floating element in our previous work; (**b**) the 2D streamline diagram in the friction sensor; (**c**) the skin friction coefficient of the flat plate; (**d**) the optimized structure of two-layer floating element; (**e**) the velocity distributions in the X direction at floating element under a total pressure of 1.5 MPa; (**f**) the velocity distributions in the Y direction at floating element.

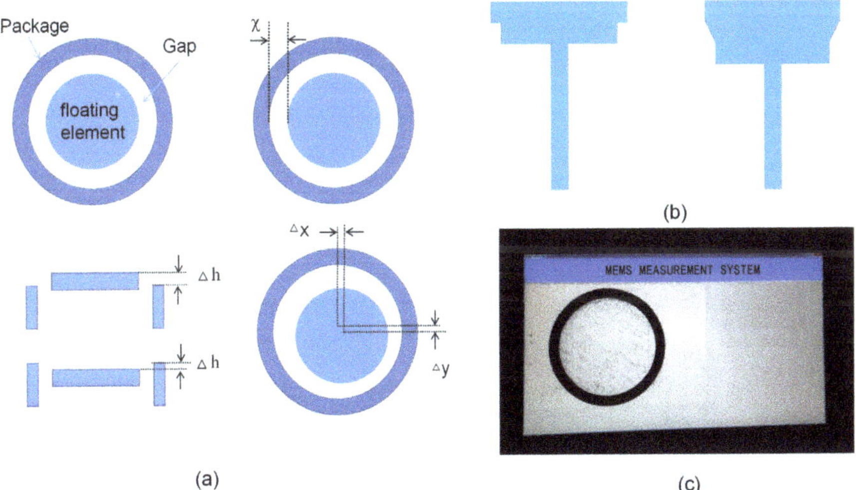

Figure 4. (**a**) The structural error of the prepared sensor; (**b**) the floating element with two-layer structure and three-layer structure; (**c**) the coaxiality measurement of the MEMS sensor.

First, the influence of two-layer and three-layer floating elements on the measurement error of friction systems under laminar and turbulent conditions is studied. The diameters of the first and second layers of the floating element are 5 mm and 0.4 mm respectively. The gap of the sensor is 100 μm. The skin friction coefficient of the sensors with different floating elements is shown in Figure 5. When the total pressure of the incoming flow is 0.5 MPa, and the levelness is 25 μm. Compared with the two-layer structure, the friction coefficient error of the three-layer structure is reduced by about 6.86% in laminar flow and increased by 0.91% in a turbulent flow. The main reason for the reduction of the error is that the shape of the floating element of the three-layer structure is obviously weaker than that of the two-layer structure. Meanwhile, the measurement error increases with the increase of the floating element bulge. It also can be seen that the floating element structure has a greater influence on the laminar flow, and the three-layer floating head structure is more conducive to reducing the measurement error.

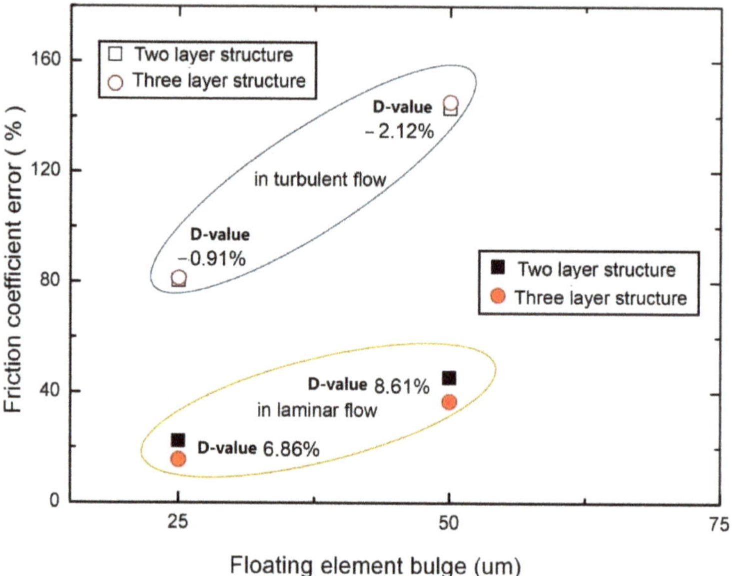

Figure 5. The friction coefficient error of the sensors with a different floating element under different flow fields.

4. Gap Optimization of the MEMS Sensor

The airflow enters the inner cavity of the sensor through the moving gap to deflect the floating element. In the numerical simulation, the friction coefficient directly read out does not take into account the influence of the incoming pressure on the floating element because the floating element cannot rotate, so it is necessary to convert the friction coefficient caused by this part of the pressure. The total friction coefficient includes the surface friction coefficient and the pressure friction coefficient.

Ideally, the floating element is perfectly flush with the measuring wall. The velocity distributions of the friction sensor with a different gap in the Y direction under laminar flow are shown in Figures 6a and 6b respectively. It can be seen that the suction effect of the moving gap is enhanced with the increase of the moving gap size, and the flow field on the measuring wall surface is also changed. The surface friction coefficient program of floating elements with different gaps is shown in Figures 6c and 6d respectively. The results show that the direct friction coefficient at the edge of the floating element in the flow direction also increases with the increase of the moving gap size, and the pressure generated by the air flow entering the friction sensor on the floating element decreases with the increase of the gap size, and the gap size has little effect on the total friction coefficient under the condition of complete flush. The total friction coefficient of the sensors with a different gap is shown in Table 2.

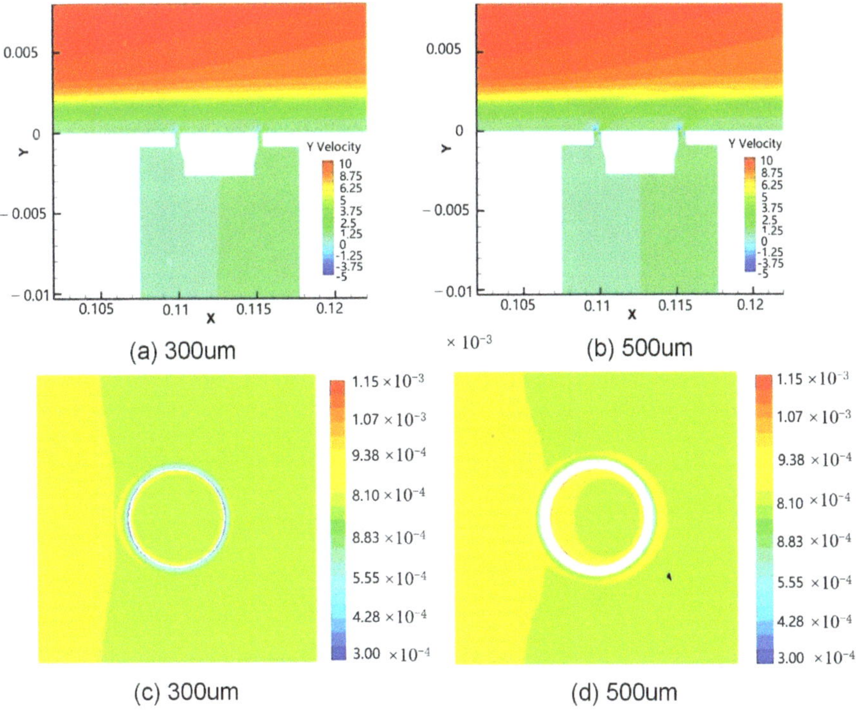

Figure 6. The velocity distributions of the friction sensor with a different gap (**a**) 300 μm and (**b**) 500 μm, and the surface friction coefficient program of the sensor with a different gap (**c**) 300 μm and (**d**) 500 μm.

Table 2. The total friction coefficient of the sensors with a different gap in laminar flow.

Gap (μm)	Misalignment (μm)	Friction Coefficient Reading Value	Pressure Friction Coefficient	Total Friction Coefficient
300	0	7.5613×10^{-4}	8.6533×10^{-6}	7.6478×10^{-4}
500	0	7.5648×10^{-4}	8.0592×10^{-6}	7.6454×10^{-4}

Although the moving g has little effect on the error of the total friction coefficient under the condition of complete alignment, it is difficult for the prepared sensor to be completely flush because the machined parts of the sensor always have errors of tens of microns. Combined with the actual assembly error of the MEMS sensor, it is assumed that the protrusion of the floating element is 50 μm. When the floating element is higher than the measuring wall, the effect of the incoming pressure on the floating element becomes obvious, which makes the pressure friction coefficient increase rapidly. At this time, the influence of the gap on the total friction coefficient error increases significantly. The effect of different gaps on the total friction coefficient in laminar flow is shown in Table 3. It can be seen that the direct friction coefficient error changes little with the increase of the moving gap and remains at about 11%. At the same time, the pressure resistance of the incoming flow has an obvious effect on the whole floating element, and the moving gap has a very significant effect on the error of the friction coefficient of the pressure resistance. The pressure friction coefficient error largely determines the total friction coefficient error (the error accounts for 88% of the total error value when the moving gap is 25 μm). Therefore, it is important to control the misalignment of the sensor with a small moving gap to reduce the friction coefficient error caused by pressure.

Table 3. Effect of different gaps on total friction coefficient in laminar flow.

Gap (μm)	Friction Coefficient Reading Value	Pressure Friction Coefficient	Total Friction Coefficient	Friction Coefficient Reading Value Error	Pressure Friction Coefficient Error	Total Friction Coefficient Error
25	8.4706×10^{-4}	6.6998×10^{-4}	1.5170×10^{-4}	12.05%	88.63%	100.68%
50	8.4608×10^{-4}	3.4693×10^{-4}	1.1930×10^{-4}	11.92%	45.89%	57.82%
75	8.4576×10^{-4}	2.8340×10^{-4}	1.1292×10^{-4}	11.88%	37.49%	49.37%
100	8.4513×10^{-4}	2.5258×10^{-4}	1.0977×10^{-4}	11.80%	33.41%	45.21%
300	8.4171×10^{-4}	1.7997×10^{-4}	1.0217×10^{-4}	11.34%	23.81%	35.15%
500	8.3676×10^{-4}	1.4434×10^{-4}	9.8110×10^{-4}	10.69%	19.09%	29.78%

The total friction coefficient error of the sensor with a different gap in laminar flow is shown in Figure 7a. It can be seen that the total friction coefficient error exceeds 100% under laminar flow when the gap of the sensor is less than 25 μm, and the total friction coefficient error quickly decreases to 45% with the gap increasing to 100 μm. As the gap increases from 100 to 500, the error only decreases from 45% to 30%, which also shows that when the gap is greater than 300, the impact of the gap on the error becomes weaker, and the impact of the misalignment on the error will be dominant at this time.

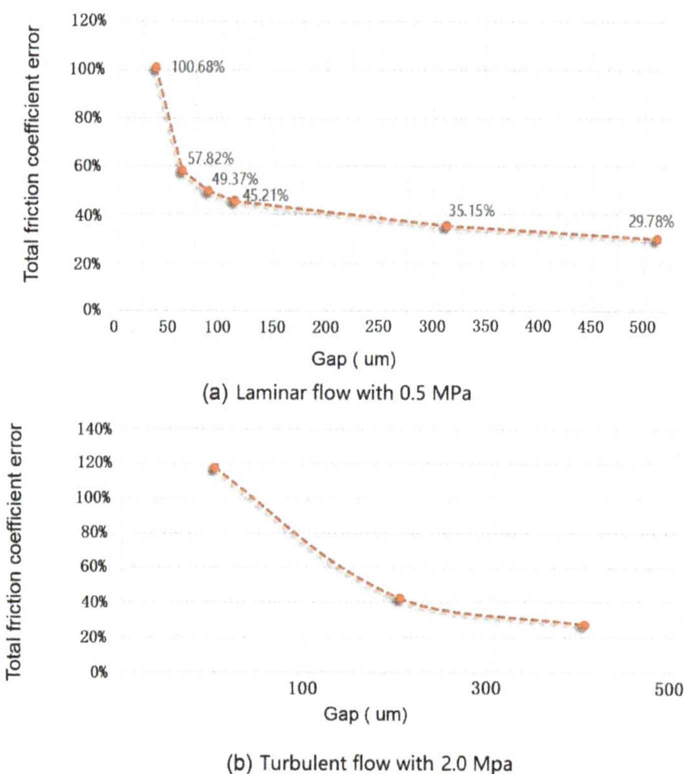

Figure 7. The total friction coefficient error of the sensor with a different gap in laminar flow (**a**) and in turbulent flow (**b**).

The total friction coefficient error of the sensor with a different gap in turbulent flow is shown in Figure 7b. The results show that the total friction coefficient error exceeds 120% under turbulent flow when the gap of the sensor is less than 100 μm, and the error decreases when the gap increases, which is the same as that under laminar flow. As a result, to reduce the measurement error of the friction coefficient, the gap of the sensor should not be less than 500 μm, and it is necessary to minimize the misalignment error at the same time.

5. Misalignment Optimization of the MEMS Sensor

The structure of MEMS skin friction is composed of a silicon differential-capacitor embedded with a floating element, signal readout circuit, and package metal shell. The machining accuracy of parts such as floating elements and package metal shells is about 10 microns. Therefore, the total misalignment error is usually in the range of −50 μm to 50 μm.

The influence of different misalignment on friction measurement error is analyzed, and the total friction coefficient error of the sensor with different misalignment in laminar flow (0.5 MPa) is shown in Figure 8. The results show that the influence of the convex of the floating element on the friction coefficient is higher than that of the concave of the floating element, and the friction measurement error decreases with the increase of the gap. Therefore, it is necessary to ensure that the floating element is lower than the measuring wall surface to avoid reverse steps affecting the flow field and increasing the measurement error. In addition, the misalignment error is reduced to 20 μm by selecting parts and controlling the adhesive thickness in practice.

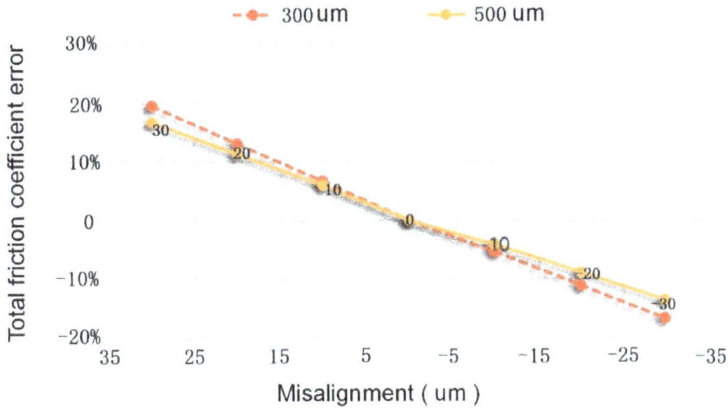

Figure 8. Influence of different misalignment on friction measurement error in laminar low.

6. Conclusions

Due to the components of the sensor including MEMS chip, floating element, signal readout circuit, and package shell being fabricated and assembled by using various processes, the structural error of the prepared sensor including the gap, misalignment, and non-coaxial between the floating element and the package is inevitable. Those structure errors of the sensor including the protrusion of the sensing element and the measurement gap will change the characteristics of the fluid, which will seriously affect the accuracy of the sensor in measuring the skin friction resistance. This paper developed a 3D numerical model of MEMS skin friction sensor in the hypersonic flow field based on CFD to research the influence of sensor-sensitive structure on wall-flow characteristics and friction perceptual characteristics. The numerical simulation method and calculation model constructed are effective, which is verified by using the Blathius solution of two-dimensional laminar flow on a flat plate, and the error is less than 0.4%. In addition, to improve the measurement accuracy of skin friction, the design criterion of the sensor-sensitive unit structure is obtained as follows:

(1)   Compared with the turbulent flow, the three-layer structure of the floating head has a greater impact on the laminar flow, and the three-layer structure of the floating head is better than the two-layer structure.

(2)   The moving gap of the floating element in the friction sensor has a great impact on the friction measurement. Under the same alignment, the friction coefficient error decreases with the increase of the moving gap, so the moving gap should be appropriately increased.

(3)   The influence of the floating element misalignment on the friction measurement is significantly reduced with the increase of the clearance. In practice, it is necessary to ensure that the floating element is lower than the measuring wall surface to avoid reverse steps affecting the flow field and increasing the measurement error.

Then, based on the above rules, the friction coefficient measurement error of the MEMS sensor with a three-layer floating element structure, a 500 μm moving gap, and the floating element misalignment between 10 μm and -20 μm is better than 10%. The future work is to prepare the sensor with special parameters and design the corresponding experimental scheme to verify the effectiveness of the model.

Author Contributions: Conceptualization, X.W. and H.G.; methodology, T.L.; Validation, H.G. and T.L.; data curation, T.L.; writing—original draft preparation, H.G.; writing—review and editing, X.W. and Y.G.; visualization, T.L. and Z.G.; project administration, H.G.; funding acquisition, H.G. and Y.G. All authors have read and agreed to the published version of the manuscript.

Funding: This research was funded in part by the Sichuan Science and Technology Program under Grant 2021YJ0105 and in part by the Science and Technology Foundation of Southwest University of Science and Technology under Grant 19zh0116, and 20zx7114.

Data Availability Statement: Data are contained within the article.

Conflicts of Interest: The authors declare no conflict of interest.

References

1.   Jiang, Z.; Farmer, K.R.; Vijay, M. A MEMS device for measurement of skin friction with capacitive sensing. In Proceedings of the Microelectromechanical Systems Conference, Interlaken, Switzerland, 21–25 January 2001; Volume 8, pp. 4–7.

2.   Mills, D.A.; Patterson, W.C.; Keane, C.; Sheplak, M. Characterization of a fully-differential capacitive wall shear stress sensor for low-speed wind tunnels. In Proceedings of the 2018 AIAA Aerospace Sciences Meeting, Kissimmee, FL, USA, 8–12 January 2018; p. 0301.

3.   Guo, H.; Wang, X.; Liu, T.; Guo, Z.; Gao, Y. MEMS Skin Friction Sensor with High Response Frequency and Large Measurement Range. *Micromachines* **2022**, *13*, 234. [CrossRef] [PubMed]

4.   Freidkes, B.; Mills, D.A.; Keane, C.; Ukeiley, L.S.; Sheplak, M. Development of a two-dimensional wall shear stress sensor for wind tunnel applications. In Proceedings of the AIAA Scitech, San Diego, CA, USA, 5–9 January 2019; p. 2045.

5.   Ghouila-Houri, C.; Talbi, A.; Viard, R.; Gallas, Q.; Garnier, E.; Molton, P.; Delva, J.; Merlen, A.; Pernod, P. MEMS high temperature gradient sensor for skin-friction measurements in highly turbulent flows. *IEEE Sens. J.* **2020**, *21*, 9749–9755. [CrossRef]

6.   Weiss, J.; Jondeau, E.; Giani, A.; Charlot, B.; Combette, P. Static and dynamic calibration of a MEMS calorimetric shear-stress sensor. *Sens. Actuators A Phys.* **2017**, *265*, 211–216. [CrossRef]

7.   Von Papen, T.; Buder, U.; Ngo, H.D.; Obermeier, E. A second generation MEMS surface fence sensor for high resolution wall shear stress measurement. *Sens. Actuators A Phys.* **2004**, *113*, 151–155. [CrossRef]

8.   Freidkes, B.R. The Design and Characterization of a MEMS-Based Dual-Axis Wall Shear Stress Sensor. Ph.D. Thesis, University of Florida, Gainesville, FL, USA, 2021.

9.   Kim, T.; Saini, A.; Kim, J.; Gopalarathnam, A.; Zhu, Y.; Palmieri, F.L.; Wohl, C.J.; Jiang, X. Piezoelectric floating element shear stress sensor for the wind tunnel flow measurement. *IEEE Trans. Ind. Electron.* **2016**, *64*, 7304–7312. [CrossRef] [PubMed]

10.   Wang, X.; Wang, N.; Xu, X.; Zhu, T.; Gao, Y. Fabrication and Hypersonic Wind Tunnel Validation of a MEMS Skin Friction Sensor Based on Visual Alignment Technology. *Sensors* **2019**, *19*, 3803. [CrossRef] [PubMed]

11.   Meritt, R.J.; Donbar, M.J.; Molinaro, J.N.; Schetz, J.A. Error source studies of direct measurement skin friction sensors. In Proceedings of the 53rd AIAA Aerospace Sciences Meeting, Kissimmee, FL, USA, 5–9 January 2015; p. 1916.

12.   Tian, Y.; Han, Y.; Yang, S.; Zhong, F.; Le, J. Investigation of fluctuating characteristics of wall shear stress in supersonic flow. *Phys. Fluids* **2019**, *31*, 125110.

MDPI AG

Grosspeteranlage 5

4052 Basel

Switzerland

Tel.: +41 61 683 77 34

Micromachines Editorial Office

E-mail: micromachines@mdpi.com

www.mdpi.com/journal/micromachines